Recent Results in Cancer Research

Fortschritte der Krebsforschung

Progrès dans les recherches sur le cancer

19

Edited by

V. G. Allfrey, New York · M. Allgöwer, Basel · K. H. Bauer, Heidelberg · I. Berenblum, Rehovoth · F. Bergel, Jersey, C. I. · J. Bernard, Paris · W. Bernhard, Villejuif N. N. Blokhin, Moskva · H. E. Bock, Tübingen · P. Bucalossi, Milano · A. V. Chaklin, Moskva · M. Chorazy, Gliwice · G. J. Cunningham, London · W. Dameshek, Boston M. Dargent, Lyon · G. Della Porta, Milano · P. Denoix, Villejuif · R. Dulbecco, La Jolla · H. Eagle, New York · R. Eker, Oslo · P. Grabar, Paris · H. Hamperl, Bonn R. J. C. Harris, London · E. Hecker, Heidelbreg · R. Herbeuval, Nancy · J. Higginson, Lyon · W. C. Hueper, Fort Myers, Florida · H. Isliker, Lausanne · D. A. Karnofsky, New York · J. Kieler, København · G. Klein, Stockholm · H. Koprowski, Philadelphia · L. G. Koss, New York · G. Martz, Zürich · G. Mathé, Villejuif · O. Mühlbock, Amsterdam · W. Nakahara, Tokyo · G. T. Pack, New York · V. R. Potter, Madison · A. B. Sabin, Cincinnati · L. Sachs, Rehovoth · E. A. Saxén, Helsinki W. Szybalski, Madison · H. Tagnon, Bruxelles · R. M. Taylor, Toronto · A. Tissières, Genève · E. Uehlinger, Zürich · R. W. Wissler, Chicago · T. Yoshida, Tokyo

Editor in chief

P. Rentchnick, Genève

Springer-Verlag Berlin · Heidelberg · New York 1968

The Cytoplasm of Hepatocytes during Carcinogenesis

*Electron- and Lightmicroscopical Investigations
of the Nitrosomorpholine-intoxicated Rat Liver*

By

Peter Bannasch

With 66 Figures

Springer-Verlag Berlin · Heidelberg · New York 1968

*Dr. Peter Bannasch, Priv.-Doz., Institute of Pathology
of the University of Würzburg, Germany*

Sponsored by the Swiss League against Cancer

ISBN-13: 978-3-642-88401-6   e-ISBN-13: 978-3-642-88399-6
DOI: 10.1007/ 978-3-642-88399-6

Softcover reprint of the hardcover 1st edition 1968

Meiner Frau

Contents

Acknowledgments

The study was presented in German to the Medical Faculty of the University of Würzburg as a part of the necessary prerequisites to obtain the venia legendi („Das Cytoplasma der Leberzelle während der Carcinogenese", Habilitationsschrift, Würzburg, 1967). It was translated by Dr. A. Schimpl, whose valuable work is gratefully acknowledged. I am also indebted to Professors H.-W. Altmann and A. Trebst and to Drs. B. Schultze and W. Thoenes for numerous suggestions and valuable criticism. The experiments were generously supported by a grant from the Deutsche Forschungsgemeinschaft. They were carried out with the expert technical assistance of Renate Beretz, Erika Dell, Ingrid Friedrich, Alfred Herbert, Elisabeth Kern, Helga Kirsch, Helga Rabenalt, Anneliese Reichert and Robin Wacker. I also want to thank Miss G. Amend for typing the manuscript.

A. Introduction

The transformation of a normal cell into a cancer cell is not a sudden but a slow continuous process which may take years. A systematic study of the morphological and structural changes that take place during this cellular transformation has only become possible since methods were developed to induce a high incidence of tumors in experimental animals. The growth of such tumors can be followed during all stages of their development. For several reasons, rat liver has proved to be particularly suitable. For more than thirty years it has been known that one can induce tumors in rat liver experimentally with many different substances. The azo dyes which were used originally (YOSHIDA, 1932; KINOSITA, 1937) have since been replaced by more potent carcinogens. The recently discovered nitrosamines, dimethylnitrosamine (MAGEE and BARNES, 1956), diethylnitrosamine (SCHMÄHL et al., 1960) and N-nitrosomorpholine (DRUCKREY et al., 1961) should be mentioned in this context. Used in the proper dose, these carcinogens lead within a few months to the formation of multicentric hepatomas in practically 100% of the experimental animals, and are therefore a very useful tool for studies of the cytogenesis of cancer (see BÜCHNER, 1961; GRUND-MANN, 1961; GRUNDMANN and SIEBURG, 1962; OEHLERT and HARTJE, 1963; BAN-NASCH and MÜLLER, 1964). The liver parenchyma as such offers a very big technical advantage because it consists of a rather homogenous cell population. Its reaction to a carcinogenic stimulus is basically uniform over large areas of the organ, as can be seen from the multicentric character of liver cell tumors. Finally the clearcut toxicological basis of chemical hepatocarcinogens, revealed by intense studies of the dose-response relationships should be emphasized (DRUCKREY and KÜPFMÜLLER, 1949; DRUCKREY and SCHMÄHL, 1962).

The above mentioned advantages are the reasons why the cytogenesis of experimental hepatomas has been extensively studied by light and electron microscopy ever since YOSHIDA's and KINOSITA's fundamental work. Morphological changes have been revealed in the course of these studies in the nucleus as well as in most cytoplasmic constituents of the intoxicated hepatocytes and the final tumor cells. The decisive question, however, as to which of these morphologically observed cellular changes are causally related to carcinogenesis and which ones are only concomitant alterations of this process has not been satisfactorily answered. We therefore undertook to reinvestigate this problem by means of comparative studies with the light and electron microscope of nitrosomorpholine intoxicated rat liver. We varied both the dose of the carcinogen and the time of action within wide limits in order to apply present day knowledge of the pharmacodynamics of hepato-carcinogenesis to the interpretation of the results obtained from morphological studies. We were particularly interested in the reversibility of toxic changes within liver cells after removal of the carcinogenic stimulus.

We shall limit the description and the discussion of our results to the cytoplasm, for only there could certain regularities be established in the morphological changes of the liver cells during carcinogenesis. There is no doubt that nitrosomorpholine also leads to considerable nuclear changes such as nuclear enlargement and mitotic anomalies. These morphological phenomena are nonspecific, however, and we consider it unlikely that they represent a basic prerequisite for carcinogenesis (BANNASCH and MÜLLER, 1964). We must stress, however, that by avoiding a description of morphological nuclear changes we do not imply any prejudice as to the primary target of a carcinogen in the cell.

B. Material and Methods

I. Animal Experiments

Inbred rats of the two genetically closely related strains BDI and BDII (DRUCKREY et al., 1962; inbred in our own laboratory) were used in all the experiments. Two to three animals, each weighing approximately 200 g at the beginning of the experiment, were kept in one cage. They were fed Altromin R ad libitum and given ordinary tap water, containing different concentrations of N-nitrosomorpholine[1] in solution. It was possible to determine the exact dose of carcinogen taken up by one set of animals by measuring the difference between the amount of drinking water supplied (20 ml per animal, per day; 6 days a week) and the remaining volume.

Eighteen out of 120 rats were used as controls. Most of the remaining animals received a 12 mg% solution of NNM, either continuously ("continuous" experiment) or over a short period of time ("stop" experiment). Two experiments were further designed to investigate the effect of lower or higher concentrations of the carcinogen (6 mg% and 20 mg% NNM solutions, respectively). Detailed experimental data are given below.

1. 6 mg% NNM-solution, continuous experiment: Fourteen BDI rats. Two animals were sacrificed after two weeks (total NNM-dose $D \sim 12$ mg/animal), 4 weeks ($D \sim 24$ mg/animal), 8 weeks ($D \sim 51$ mg/animal), 14 weeks ($D \sim 84$ mg/animal), 22 weeks ($D \sim 133$ mg/animal), 34 weeks ($D \sim 213$ mg/animal) and 37 weeks ($D \sim 227$ mg/animal), respectively. In this case samples were studied under the light microscope only.

2. 12 mg% NNM-solution, continuous experiment: Twenty-six BDI rats. Three rats were sacrificed after 2 weeks ($D \sim 27$ mg/animal), 4 weeks ($D \sim 47$ mg/animal), 8 weeks ($D \sim 99$ mg/animal), 13 weeks ($D \sim 146$ mg/animal), 17 weeks ($D \sim 197$ mg per animal), 19 weeks ($D \sim 223$ mg/animal), 27 weeks ($D \sim 306$ mg/animal). Five of the rats died between the 25th and the 28th week ($D \sim 303$—343 mg/animal). Results of

---

[1] We gratefully acknowledge the help of Dr. R. PREUSSMANN (Forschergruppe Präventivmedizin am Max-Planck-Institut für Immunbiologie, Freiburg i. Br., Leiter: Prof. Dr. H. DRUCKREY), who provided us with the N-nitrosomorpholine.

this experiment obtained by studies with the light microscope, have already been published in detail (BANNASCH and MÜLLER, 1964). In the meantime the whole series was subjected to electron microscope investigations for comparison.

3. 12 mg%₀ NNM-solution, stop experiment: Fourty-four BDI rats. Three animals were sacrificed after 2 weeks (D∼25 mg/animal), 4 weeks (D∼54 mg/animal), 8 weeks (D∼127 mg/animal), 11 weeks (D∼145 mg/animal) and 12 weeks (D∼167 mg per animal), respectively. Administration of the carcinogen was stopped after 12 weeks, when a total of 130—170 mg had been taken up by each animal. Three to six rats were killed 4 days and 2, 4, 6, 8 and 10 weeks, respectively, after removal of the carcinogenic stimulus. Four of the rats died, one during the carcinogenic treatment (12 weeks, D∼161 mg), the others, 5, 8 and 9 weeks after removal of the chemical from the drinking water. The livers of all animals were examined under the light microscope. Animals which had been killed 4 days, 4 weeks and 8 weeks after the end of the treatment were chosen for investigation under the electron microscope.

4. 20 mg%₀ NNM-solution, stop experiment: Eighteen BDI rats. All animals received the carcinogen with their drinking water over a period of 7 weeks. The total amount of carcinogen taken up by each animal varied from 141 to 166 mg, which corresponds approximately to the doses of the 12 weeks treatment with a 12 mg%₀ NNM solution (see above). One set consisting of three animals was sacrificed immediately after the end of the carcinogen uptake, the rest 2, 4, 5, 10 and 21 weeks later. The livers of all animals were again examined under the light microscope. Those rats killed 21 weeks after the removal of the carcinogen were subjected to studies under the electron microscope.

II. Techniques of Light Microscopy

With the exception of the animals that died spontaneously, the liver tissue was removed under ether anaesthesia, fixed in a 4% formaldehyde-solution, Carnoy's and Zenker's fixative. The samples were embedded in paraffin. In those cases where glycogen was to be revealed by Best's Carmine (counterstain haemalaun) or by the PAS reaction (Tri-PAS), a 70% alcohol was used instead of the conventional water-bath to avoid a partial elution of glycogen.

Other stains used in the experiments: hematoxylin-eosin, Cresyl violet, v. Gieson, Goldner, Feulgen. Fat was stained with scarlet red, using frozen sections. Several liver lobes of one animal were used in each case.

III. Techniques of Electron Microscopy

One mm³ sections were made from tissue that had been obtained from living animals. They were fixed in a 1% OsO₄-solution, buffered at pH 7,3. The samples were rinsed in distilled water for two hours; the tissue was dehydrated in a graded solution of ethanol (50, 70, 90 and 100%). The sections were embedded in Vestopal-W (initiator 1%, activator 0,5%) (RYTER and KELLENBERGER, 1958), using styrene as an intermediate (KURTZ, 1963). Polymerization was carried out at 47 °C for 24 hours, followed by 24 hours at 60 °C. Thin sections were cut with a Porter-Blum microtome and stained for 10 minutes with lead hydroxide (KARNOVSKY, 1961) or for 1—2 hours with uranyl acetate (WATSON, 1959). Siemens Elmiskop I (60 KV, objective aperture 30 μ). Ilford plates N 60. Tepa developer (Agfa).

Sections of 1—2 μ were cut from each block for comparative investigations with the light microscope. They were stained with a saturated solution of toluidine blue (GAUTIER, 1960), a concentrated solution of Giemsa (THOENES, 1960) or PAS. Thus an exact comparison of changes by light and electron microscopy was possible.

C. Results

The structure of normal liver cells has been described by many authors and has recently been reviewed in several articles (ROUILLER and JEZEQUEL, 1963; DAVID, 1964; COSSEL, 1964; BRUNI and PORTER, 1965; and others). Therefore, we need not describe the livers of our control animals in detail. There is one fact, however, which will be important for the discussion of our results, namely the glycogen content of liver cells. Under the experimental conditions at hand, one can observe a distinct storage of glycogen throughout the liver parenchyma of the control animals. The size of the deposits, however, varies considerably from cell to cell. These variations in glycogen content are always connected with variations in the fine structure of the cytoplasm which have already been described in detail for other controls (THOENES and BANNASCH, 1962).

I. The Precancerous Phase

For animals which are continuously fed a 12 mg% NNM-solution, earlier investigations with the light microscope have revealed that the changes in the cytoplasm of liver cells during the precancerous phase at the center of the lobules are basically different from those at the periphery (BANNASCH and MÜLLER, 1964). To describe these phenomena as "acinuscentral" and "acinusperipheral" in accordance with KIERNAN's strictly morphological classic concept of liver lobules (KIERNAN, 1833) is, however, not fully satisfactory. At least during the first weeks of the experiment, the histotoxic pattern shows a definite correlation with the different zones of vascula-

Fig. 1 a—d. Loss of glycogen from the liver parenchyma as a function of carcinogen concentration. a 6 mg% NNM-solution, 8 weeks: very few glycogen-free hepatocytes in the immidiate vicinity of the central vein. b 12 mg% NNM-solution, 8 weeks: total loss of glycogen in the third zone of the "functional" liver acinus as defined by RAPPAPORT; pronounced storage of glycogen in the first and second zones. c 20 mg% NNM-solution, 7 weeks: advanced loss of glycogen throughout the lobule, between the glycogen-free cells some epithelia with extensive glycogen storage. d 20 mg% NNM-solution for 7 weeks, followed by plain water for 6 weeks: pronounced glycogen storage throughout the parenchyma. All figures: Tri-PAS. 120:1

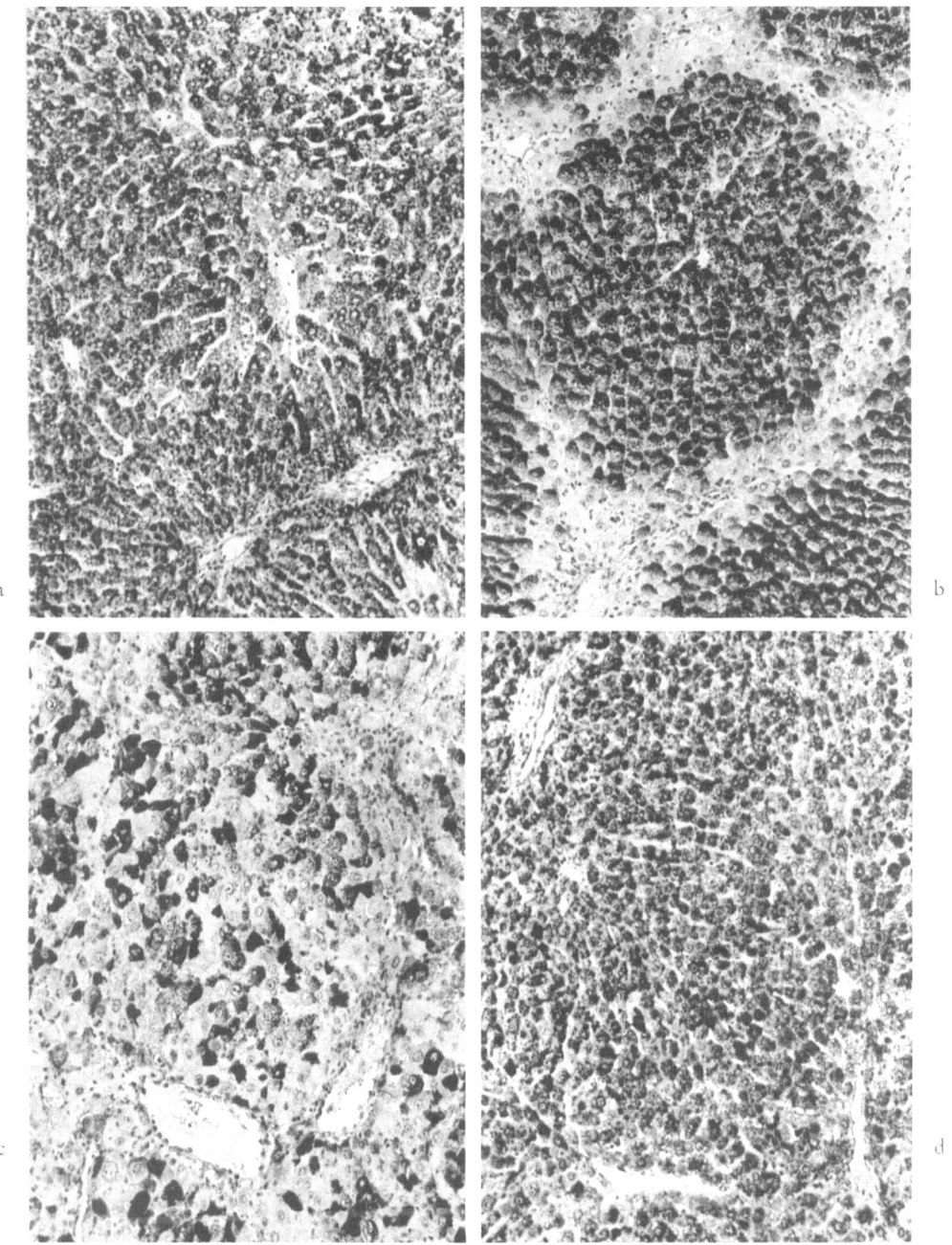

Fig. 1 a—d

rization, described by RAPPAPORT et al. (1954), which are collectively attributed to
a "functional" liver acinus. Unfortunately, the concept of RAPPAPORT cannot be used
more effectively than the old one to order the present findings in a clear fashion.
The main reason for the difficulty is that a cirrhotic reorganization of the liver
parenchyma occurs in the later stages of the experiment, leading to a considerable
disturbance of the normal blood flow. In addition, there is a remarkable shift in the
histotoxic pattern depending on the concentrations of the carcinogen. As KIERNAN's
system has the advantage of being generally known, we shall distinguish between an
"acinuscentral" and an "acinusperipheral" cytotoxic pattern, meaning thereby that
the cellular changes occur preferentially in the center or at the periphery of the liver
lobules as defined in classic terms. Using practical examples, we shall later discuss
certain overlappings between the two cytotoxic patterns.

1. The Acinuscentral Cytotoxic Pattern

a) Light Microscopy

Cytologic Characteristics

Two typical changes in the cytoplasm characterize the central cytotoxic pattern:
first, a reduction or *loss of glycogen* (Fig. 1) and second, a *disaggregation of the
basophilic bodies,* typical of the normal ergastoplasm, resulting in a diffuse cytoplasmic
basophilia (Figs. 2 and 3 b). Both cytoplasmic changes are intimately related to each
other and always occur together. The diffuse cytoplasmic basophilia may be more or
less pronounced. In some cells the cytoplasm even shows a distinct eosinophilia and
a fine granulation. The epithelia in the center of the lobules often seem to be reduced
in size. In many instances they undergo *coagulation necrosis.*

Histogenesis of the Alterations as a Function of Dose and Time

The above mentioned cytoplasmic changes usually take place in well defined areas
of the liver lobules. Their localization and size strongly depend on the concentration
of the hepatotoxin employed.

In animals which received a *12 mg⁰/₀ NNM-solution* one can observe that the
glycogen-free area ("glycogen-free" stands collectively for all the acinuscentral
changes of the parenchyma) coincides fairly well with the outer parts of the third
zone of the "functional liver acinus" as defined by RAPPAPORT. In other words, one

---

Fig. 2. In the center of the liver lobule (at the right) loss of glycogen, diffuse cytoplasmic
basophilia ("chromatolysis"), partial disappearance of the parenchyma and mesenchymal proliferation. In the intermediary and peripheral zones of the lobule (at the left) enhanced glycogen
storage (big clear cells). 12 mg⁰/₀ NNM-solution, 14 days. HE. 150:1

Fig. 3. Different behaviour of the basophilic bodies in hepatocytes of the acinusperipheral
(a) and the acinuscentral (b) cytotoxic pattern. a "Dislocation" of integral basophilic bodies
towards peripheral regions of the cytoplasm by excessively accumulated glycogen which in
this case is eluted. b "Disorganization" of the basophilic bodies entailing a "diffuse cytoplasmic
basophilia" ("chromatolysis") and concomitant loss of glycogen. 12 mg⁰/₀ NNM-solution,
14 days. Cresyl violet. 1900:1

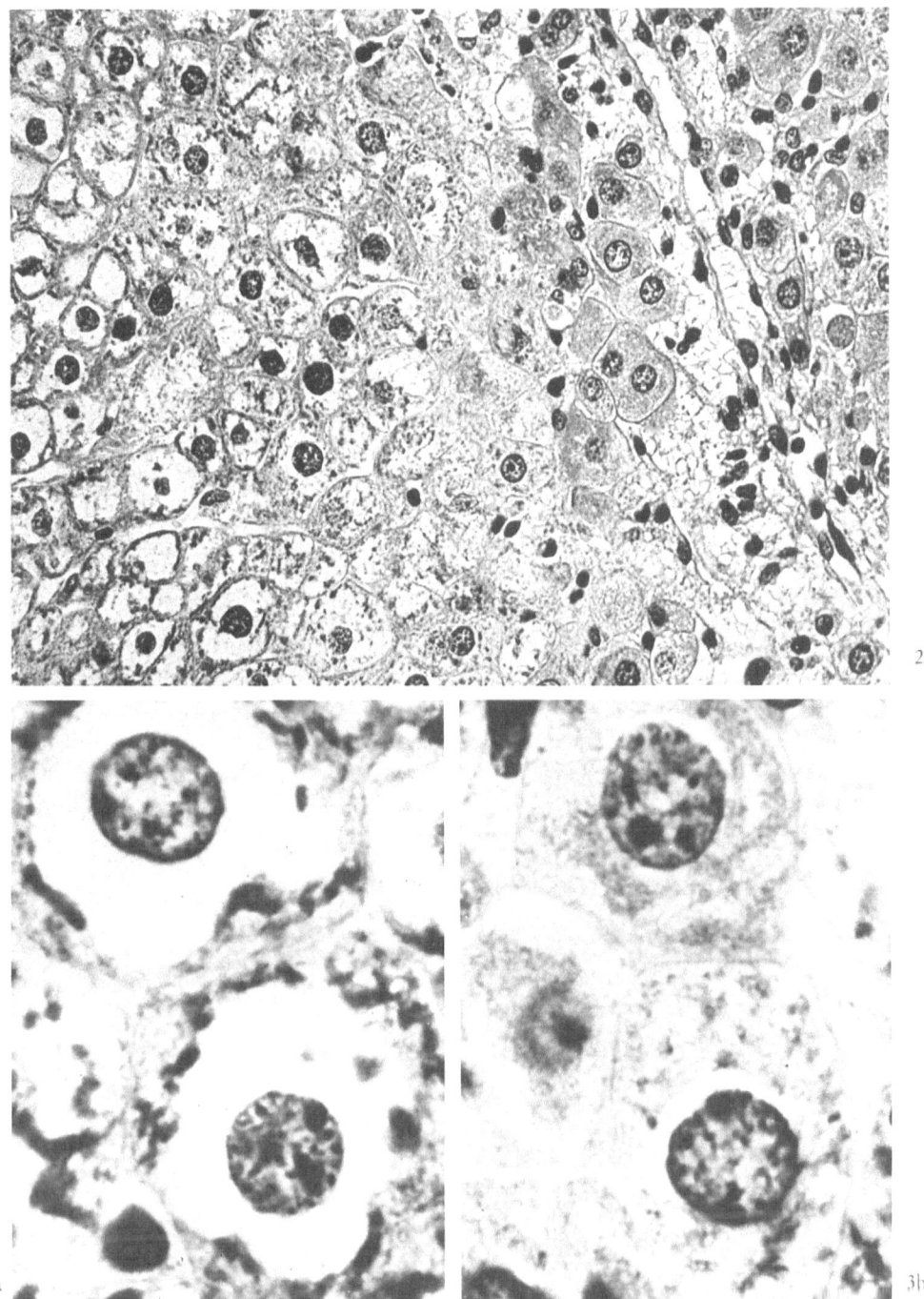

Figs. 2 and 3

finds the glycogen-free cells mainly in the center of the lobule (1—2 cell layers surrounding the central vein) and in small rays leading to the periportal regions (Fig. 1 b). This histotoxic pattern develops fully within two weeks after the beginning of the experiment and persists as long as the carcinogen is administered. The glycogen-free zones never exceed two additional cell layers, even if the toxic treatment is extended over several weeks. The number of necrotic cells also remains more or less constant throughout the entire period of carcinogen administration. In the later stages of the continuous experiment, the necroses lead to the development of liver cirrhosis in most of the experimental animals. Furthermore, there is often an extensive proliferation of bile ducts.

When the carcinogen is removed from the diet after 12 weeks *(stop experiment)*, the acinuscentral epithelia usually restitute their glycogen deposits within two weeks, and the typical basophilic bodies are formed again. Necrotic cells are very rare. If at the end of the carcinogen treatment cirrhotic or precirrhotic changes have already developed in the liver, they remain practically unchanged during the following weeks.

If a *6 mg⁰/o NNM-solution* is given continuously (up to 37 weeks) one finds throughout the experiment only rarely glycogen free, diffuse basophilic liver epithelia in the vicinity of the central vein (Fig. 1 a). The incidence of necrosis is extremely low. Therefore liver cirrhosis is never observed.

If a *20 mg⁰/o NNM-solution* is used, however, glycogen depletion and the concomitant cytoplasmic changes during carcinogen administration spread from the third to the second and in some cases even to the first zone of the "functional" liver acinus. Thus almost all of the liver lobule is affected (Fig. 1 c). Necrotic cells are observed much more frequently than in the experiment where a 12 mg⁰/o solution was used. As a consequence, liver cirrhosis develops faster and is more pronounced. After removal of the carcinogenic stimulus (7 weeks after the beginning of the experiment) the glycogen deposits throughout the parenchyma are restored (Fig. 1 d) and the ergastoplasmic bodies are formed again.

b) Electron Microscopy

Glycogen and Endoplasmic Reticulum (ER)

As in the light microscopic observations, one witnesses a total *disappearance of the glycogen* zones under the electron microscope in the acinuscentral epithelia. The agranular reticulum (AR), usually localized at the periphery or at the center of the glycogen zones, is now gathered in certain areas of the cytoplasm. In the case of a strictly central cytotoxic pattern one cannot observe a considerable enhancement of these cytoplasmic membranes. (We shall discuss later the case of glycogen-poor or glycogen-depleted cells that do show a marked increase in smooth ER. We ascribe these cells to the peripheral cytotoxic pattern.) If the section are stained with lead hydroxide one can observe in some cells single glycogen particles or glycogen rosettes between the membranes of the AR. Other cells are completely devoid of glycogen.

The ergastoplasm, that is the rough or granular reticulum (GR), is subject to striking changes (Fig. 4). The rough cisternae are no longer organized parallelly in groups but are distributed evenly throughout the cell. An *ergastoplasm disorganization* of this kind has been described by many authors as a consequence of various

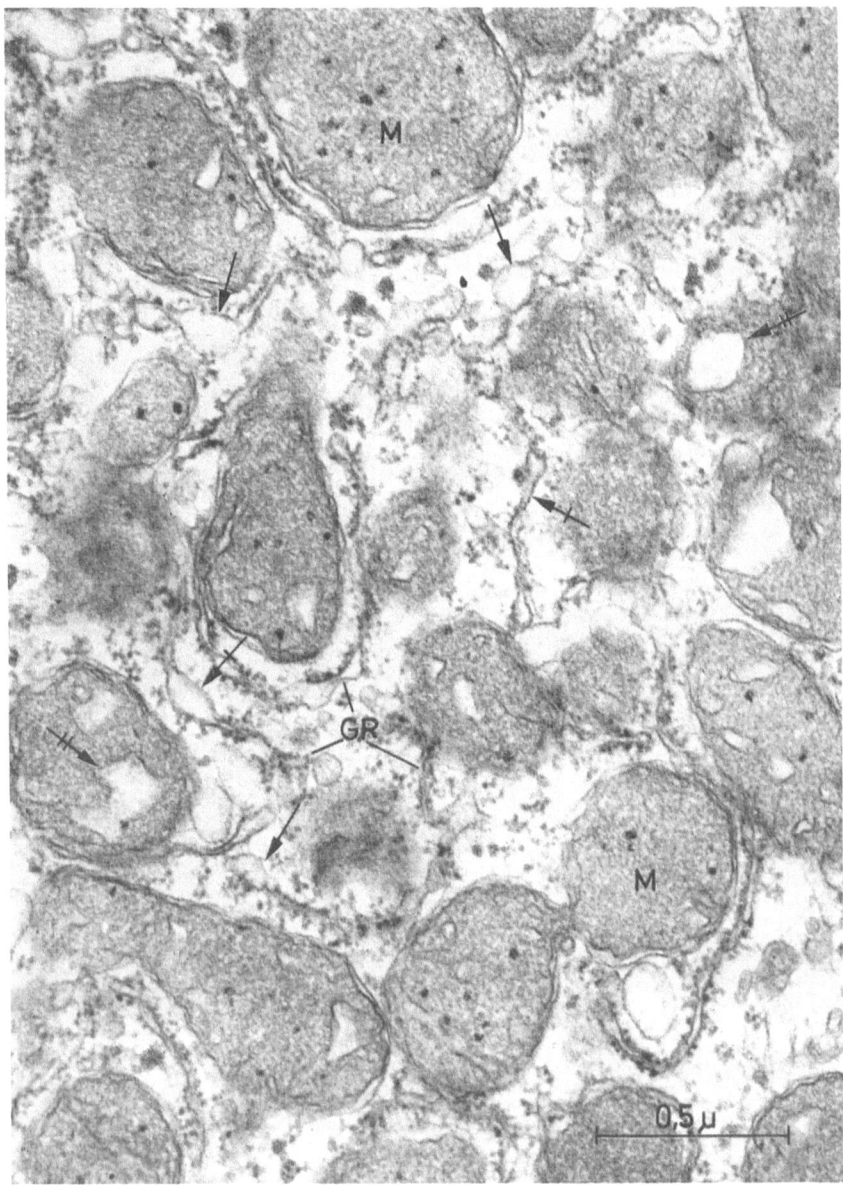

Fig. 4. Portion of a hepatocyte of the acinuscentral cytotoxic pattern. Advanced disorganization of the ergastoplasm (granular reticulum: GR). The granular profiles are scattered throughout the cytoplasm and are in close contact with mitochondria (M). Here and there the cisternae are dilated (↓) and have lost their ribosomes (†). Mitochondrial cristae are rarefied; the remaining cristae are often markedly dilated (‡). 12 mg⁰/o NNM-solution, 14 days Lead hydroxide. 52 000:1

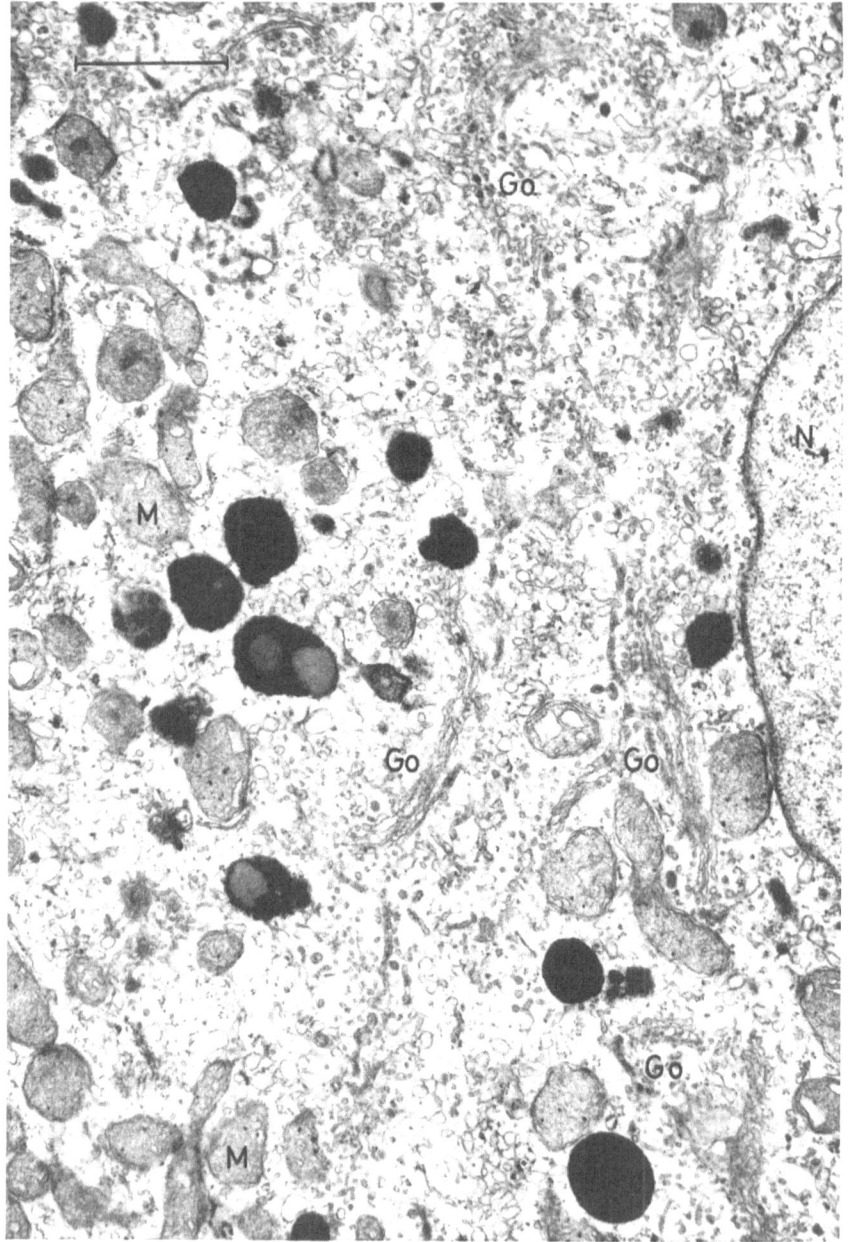

Fig. 5. Portion of a hepatocyte of the acinuscentral cytotoxic pattern. Loss of glycogen. "Vesicular transformation" of almost the entire ER. Within large areas of vesicles several Golgi-complexes (Go). In their vicinity one frequently observes "peribiliar dense bodies". Rarefaction of cristae mitochondriales (M). Nucleus (N). 12 mg% NNM-solution, 14 days. Lead hydroxide. 21 000:1

factors affecting the liver, such as prolonged starvation (BERNHARD et al., 1952; FAWCETT, 1955; and others) or intoxications (OBERLING and ROUILLER, 1956; cf. LAFONTAINE and ALLARD, 1964). It was realized earlier that this phenomenon corresponds to the diffuse basophilia which can be observed under the light microscope. As in the case of thioacetamide intoxication (THOENES and BANNASCH, 1962) one can distinguish between several degrees of disorganization of the ergastoplasm when rat liver is exposed to NNM. At least at the beginning one can observe in many cells a mere displacement of the ergastoplasmic cisternae, while the ribosomes still remain attached (Fig. 6). The rough cisternae enter into close contact with the mitochondria, which they encircle or partially surround. As the disorganization of the ergastoplasm proceeds, one observes additional changes of the cisternae (Fig. 4). The rough profiles are now partly depleted of ribosomes and swollen at various spots. Frequently, the profiles are shortened. Occasionally one can find circular membrane profiles which in most cases carry only a few ribosomes or none at all. Eventually the entire ER of the cell may be transformed into small vesicles, some rough, some smooth (Fig. 5). In severe cases, this process obviously also affects the smooth ER. We reach this conclusion not only from the appearance of smooth vesicles, which could also be derived from ergasto-plasmic cisternae having lost their ribosomes, but also from the fact that in addition to the usual elongated rough profiles the typical tubular-vesicular smooth ER is practically absent too. In most cases the small vesicles are spread irregularly throughout the cytoplasm; at some points large vesicular areas are formed. It is noteworthy that the smooth vesicles often contain some osmiophilic material.

Within the vesicular areas one can sometimes observe a *large number of Golgi complexes* (Fig. 5). They consist of three or four elongated parallelly stacked cisternae and chains of vesicles. Both cisternae and vesicles often contain a similar dense osmiophilic material, as can be observed in the vesicles which are distributed irregularly over the cytoplasm. The peribiliar Golgi-complexes of acinuscentral epithelia generally show their usual structure.

The Chondriom

The appearance of the mitochondria changes from cell to cell. Some of the glycogen-free epithelia show a chondriom which seems to be morphologically unchan-ged, even after weeks of intoxication. On the other hand, numerous cells contain mitochondria which already show more or less pronounced *structural changes* within two weeks after the beginning of carcinogen administration (Figs. 4—20). At the same time one finds in many cases an *enlargement of the chondriom:* the entire cytoplasm is densely populated with mitochondria (Figs. 6 and 7). Concerning the architecture of the mitochondria one has to differentiate between changes of either the shape or the inner structure. They do not always appear together. Within one cell, however, all the mitochondria usually show the same changes. If one takes into account both the structural changes and the number of mitochondria, one can differentiate between several characteristic states of the chondriom. There are *three basic types:*

In the *first* case, the size of the chondriom corresponds approximately to that of a normal liver cell. The shape of the mitochondria remains by and large unchanged. The inner structure, however, shows marked differences (Figs. 4 and 5): the number of cristae mitochondriales is more or less reduced; in the remaining cristae one observes an enlargement of the intracristal space, which may even look like a bag, filled with finely

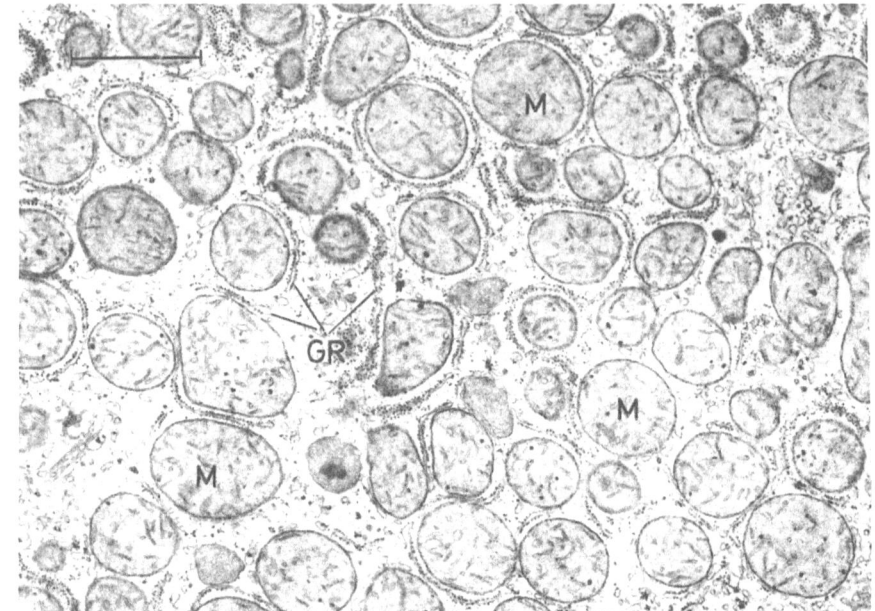

Fig. 6. Portion of a hepatocyte of the acinuscentral cytotoxic pattern. Dense population of mostly rounded mitochondria (M). Cristae cut obliquely usually show round or angular profiles. Weak disorganization of ergastoplasm: the granular profiles (GR) are scattered and in close contact with mitochondria (M). 12 mg⁰/o NNM-solution, 8 weeks. Lead hydroxide. 18 000:1

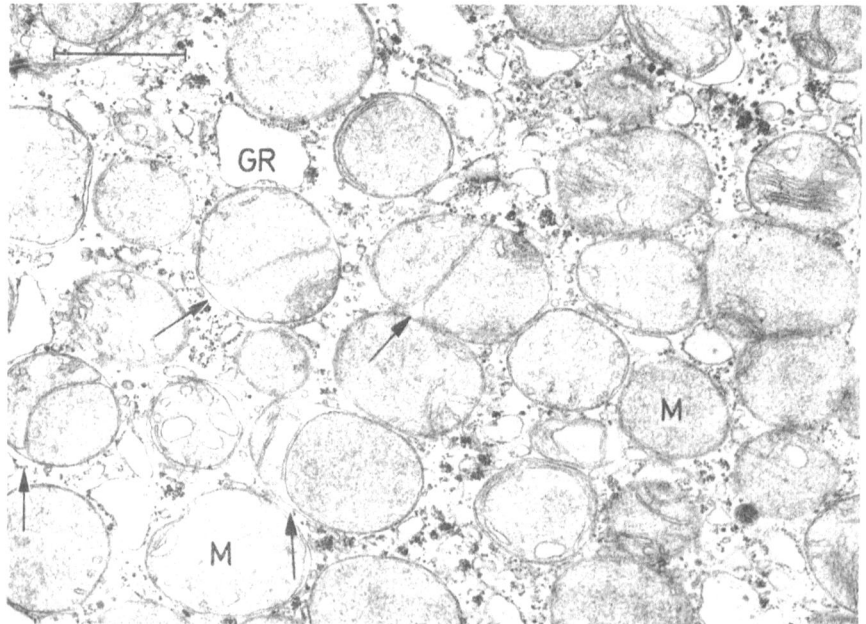

Fig. 7. Portion of a hepatocyte of the acinuscentral cytotoxic pattern. Dense population of enlarged and round mitochondria (M). Pronounced rarefaction or deformation of cristae. Many transverse partitions (↓). Advanced disorganization of the ergastoplasm together with a marked dilatation of the granular cisternae (GR) mostly containing a finely flocculated material. 12 mg⁰/o NNM-solution, 8 weeks. Lead hydroxide. 18 000:1

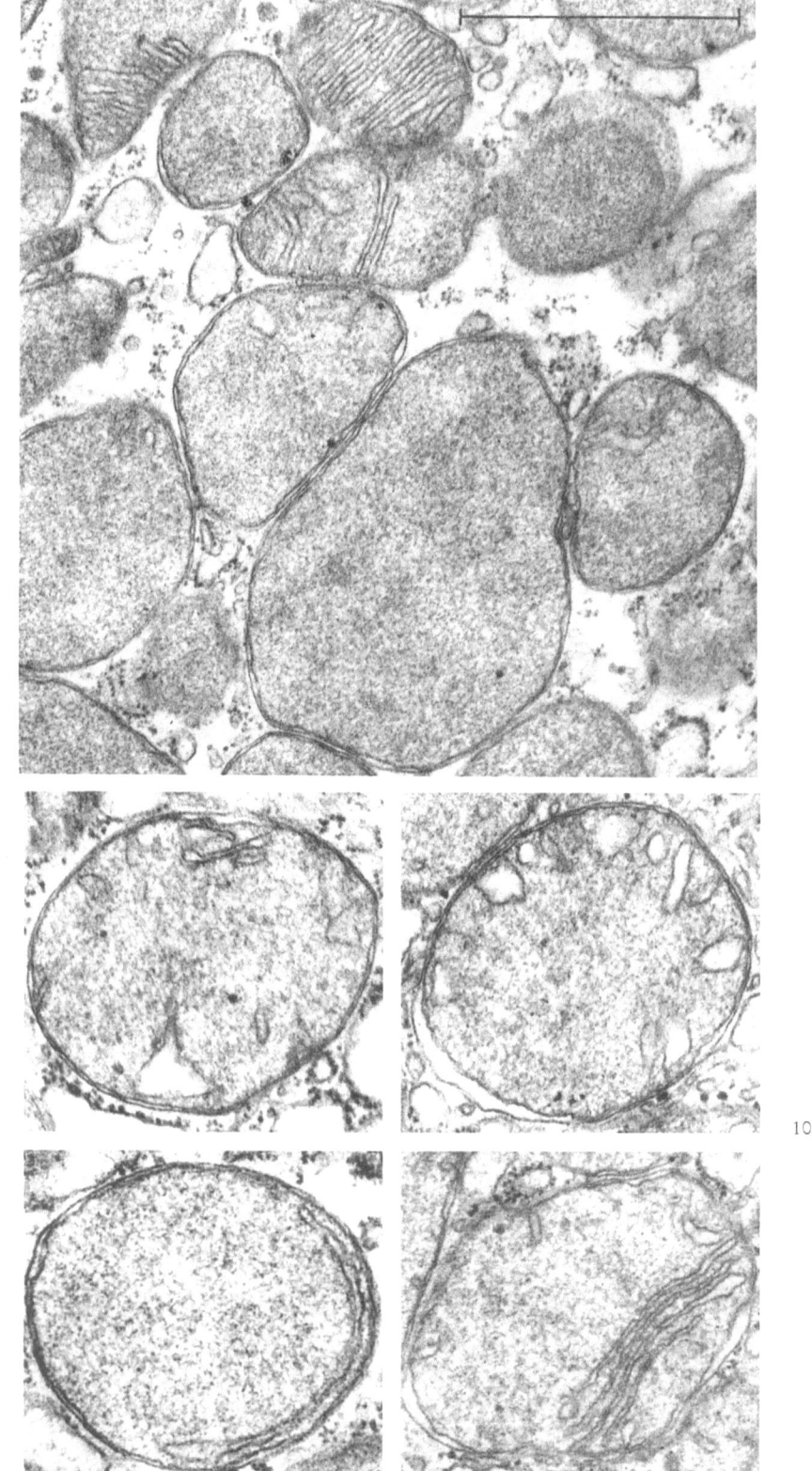

Figs. 8—12 (Legends see p. 14)

flocculated material. In some cases, the dilated intracristal space seems to be optically empty. The ergastoplasm of such cells usually shows signs of advanced disorganization.

The *second* stage is mainly characterized by an increase in the number of mitochondria (Fig. 6). Individual mitochondria show insignificant structural modifications, which one only realises, if one looks at the entire chondriom. It consists of a rather homogeneous population of round mitochondria, which look like circular or oval discs upon sectioning. Only in a very few cases does one encounter elongated and slender profiles in the plane of section. All mitochondria contain a large number of cristae in a rather irregular arrangement. Transverse sections show them to possess a small circular or triangular profile (see REVEL et al., 1963). The insertion at the inner membrane of the mitochondrial envelope can only rarely be revealed. The second state of the chondriom may still be considered a modification of that usually observed. This is corroborated by the fact that the other cytoplasmic components of such cells do not show definite lesions either. The disorganization of the ergastoplasm proceeds only to a very small degree, merely characterized by a displacement of the rough membranes thereby entering into close contact with mitochondria, which they more or less envelop.

There are all possible intermediate stages leading from the above-mentioned states of the chondriom to a *third* type (Figs. 7—20). The latter is characterized by an increase in the mitochondrial population and a net pathological structural transformation of the mitochondria. This third type may thus be considered a combination of the two others described earlier. The overall shapes of individual mitochondria partially correspond to those found with the second type of mitochondrial changes: on the average they are, however, noticeably larger. They may be very densely packed and their contours are, therefore, sometimes flattened or slightly concave. For this reason one can rarely find purely circular or oval transections. Very seldom one may encounter sections of irregular and unusually long mitochondria.

All these changes in the overall shape, characteristic for the third type, are concomitant with various transformations of the inner structure of the mitochondria which affect the cristae as well as the envelope and the matrix. A considerable *loss of cristae mitochondriales* is common in almost all mitochondria. This effect may be so pronounced that one no longer finds a single crista in the plane of section (Figs. 8, 17 and 18). The two leaflets of the mitochondrial envelope, however, remain almost invariably intact. The matrix of the mitochondriae devoid of cristae is always homogenous and finely granulated; in some cases it is rather loosely structured, in others, however, remarkably densely packed. The osmiophilic particles which are normally embedded in the matrix are usually completely absent. Thus, the changes of the mitochondria affect the structure to such an extent that the individual mitochondrion can barely be recognized as such. There is no doubt, however, that the observed

---

Fig. 8—12. Mitochondria from hepatocytes of the acinuscentral cytotoxic pattern (12 mg%/o NNM-solution, 8 weeks, lead hydroxide). 8: All possible transitions leading from "crista-rich" to "crista-less" mitochondria. Homogeneous densification of the matrix. Nearly complete loss of intramitochondrial granules. 37 000:1. 9—10: Rarefaction, deformation and dilatation of mitochondrial cristae. 44 000:1. 11: Far-reaching rarefaction of mitochondrial cristae. A remaining crista arranged parallelly to the mitochondrial envelope. 54 000:1. 12: Elongated and parallelly stacked cristae in addition to short cristal remnants. 45 000:1

organelles, devoid of cristae, actually originate from mitochondria, since one can find within a single cell all possible transitions between the cristae-less mitochondria and those still containing some cristae and even cristae-rich ones (Fig. 8).

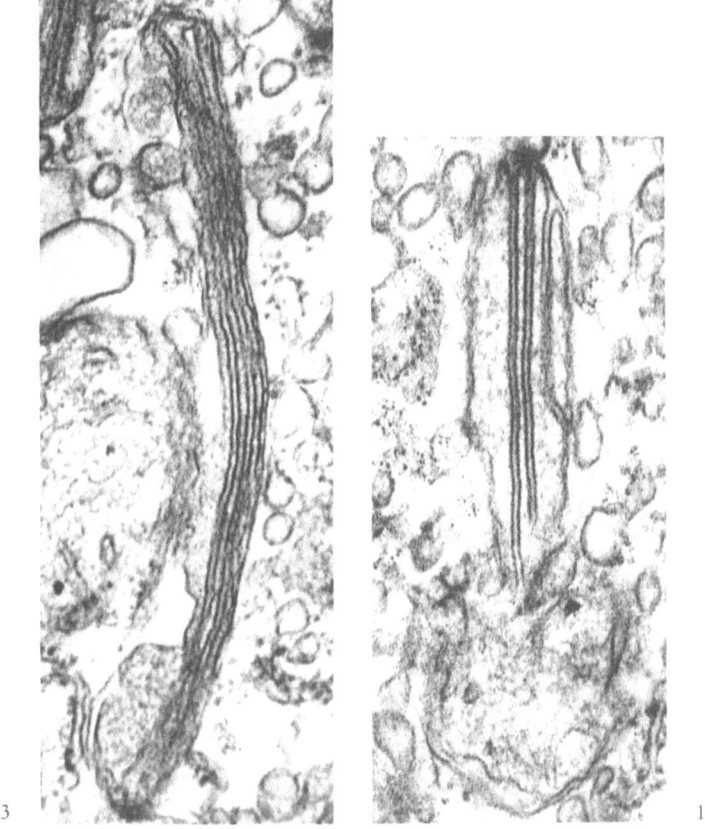

Fig. 13 and 14. Mitochondria from hepatocytes of the acinuscentral cytotoxic pattern. In the interior of the mitochondria densely packed, long cristal profiles arranged longitudinally. Note the insertion of one crista into the inner leaflet of the mitochondrial envelope in Fig. 14. 12 mg⁰/o NNM-solution. Lead hydroxide. 140 000:1

The profiles of the remaining cristae may be extremely long or very short and are frequently deformed (Figs. 7—14). They are mostly confined to the peripheral areas of the mitochondrion, thus giving the impression of having been pushed towards the envelope. The matrix usually seems to be somewhat denser in areas with remaining cristae than in the cristae-less ones. The long cristae are usually stacked together (Figs. 8, 12, 13 and 14). They mostly traverse the entire mitochondrion and show straight, wavy or sinuous paths. In the last case they usually run parallel or inverse to the mitochondrial envelope. In any case, one can find the long cristae in almost any possible position. In rare instances one finds strangely deformed long mitochondria, in which the interior is almost completely filled with densely packed membrane profiles arranged longitudinally (Figs. 13 and 14). Cuts which reveal the attachment of the membranes to the mitochondrial envelope clearly document that these profiles are in

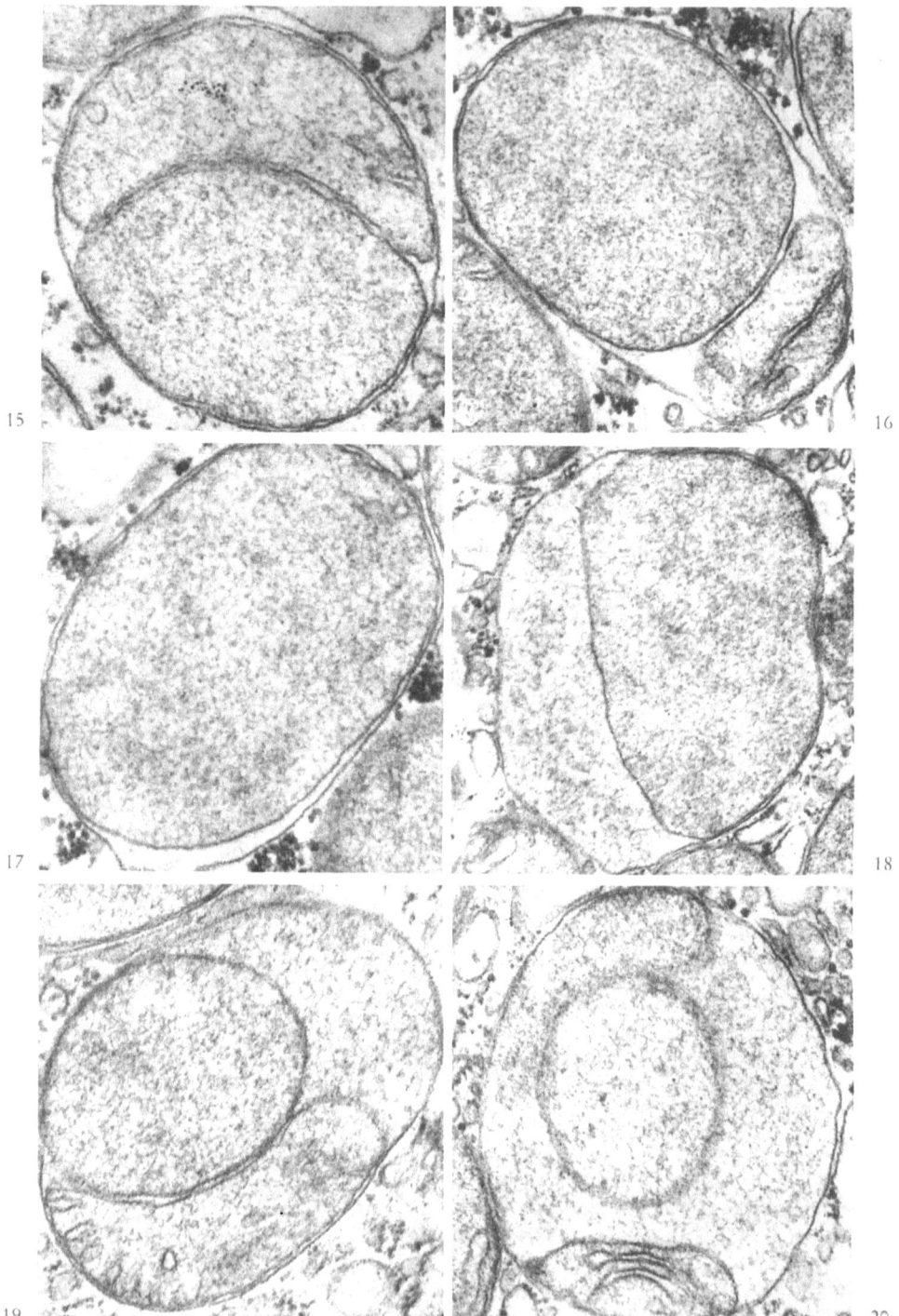

Figs. 15—20

fact derived from cristae mitochondriales which, however, have undergone severe alterations (Fig. 14). The majority of the short cristal profiles are confined to the vicinity of the mitochondrial envelope. They are frequently swollen into vesicles, in which case they contain a finely flocculated material, or else seem to be optically empty. Sometimes, one may encounter individual short profiles embedded in large matrix areas. Although in many cases one can clearly observe the insertion of the inner membrane profiles into the envelope of the mitochondrion it is questionable whether all those membranes which do not show any attachment within the plane of section are still connected to the envelope. A clearcut answer could only be obtained by means of serial sectioning.

In addition to the above mentioned changes, one observes in some mitochondriae the formation of a *diaphragm-like transverse partition* (Figs. 15 and 16). It consists of two membranes which are continuous at both sides with the inner leaflet of the mitochondrial envelope. They are separated by a small space which in itself communicates with the interspace of the envelope. As these partitions always consist of two continuous membranes one must conclude that they separate the inner compartment of the mitochondrion into two distinct chambers.

It has already been mentioned that the two leaflets of the mitochondrial envelope almost always remain intact, even in the case of a far-reaching reduction of the number of cristae. Only in a single case could we observe a rupture of the inner leaflet of the envelope. One does, however, find another peculiar *alteration of the envelope*, namely a more or less pronounced widening of some parts of the inter-leaflet space, which becomes filled with a homogeneous finely flocculated material (Figs. 17, 18 and 19). In some cases, however, the widened gap seems to be optically empty. We have never observed the insertion of cristae at the gaps of the inter-leaflet space. Parts of the mitochondrial envelope which are not dilated do not show any noticeable changes. Here, one can still find points of attachment of individual cristae; they may be swollen, but are also observed to be very narrow. Swelling of the inter-leaflet space may be so excessive that cross-sections through the dilated part of the envelope may be almost as extended as those of the entire inner compartment of the mitochondrion (Fig. 18). If the mitochondrion contains a transverse partition of the sort described above, the space between its membranes often undergoes a change in the same direction as the interleaflet space of the envelope. The two membranes of the partition are forced apart, whereby a wide communication ensues between the enlarged interspace of the partition and the dilated part of the envelope. This effect may be so pronounced in some cases that the two chambers of the inner compartment are pushed toward the opposite poles of the mitochondrion (Fig. 20). These observations do not only prove the communication between the interspaces of envelope and partition, they also clearly document that the two leaflets of the transverse partition are actually closed.

Fig. 15—20. Mitochondria from hepatocytes of the acinuscentral cytotoxic pattern 12 mg⁰/o NNM-solution, 8 weeks, lead hydroxide). 15 and 16: Marked loss of cristae mitochondriales. Separation of the inner compartment of the mitochondria by a transverse partition which is at both sides continuous with the inner membrane of the envelope. 52 000:1 and 54 000:1. 17 and 18: "Crista-less" mitochondria with a partial amplification of the interleaflet space of the envelope. 45 000:1 and 41 500:1. 19 and 20: Amplification of the envelope and the interspace of transverse partitions leading to a displacement of two separate chambers towards opposite poles of the organelle (20). 61 000:1 and 53 000:1

All the mitochondrial changes just described, which are characteristic of the third type of mitochondrial alterations, are generally associated with an advanced disorganization of the ergastoplasm (Fig. 7). It is remarkable, however, that one often finds in the immediate vicinity of the cristae-poor or cristae-less mitochondria rather large ergastoplasmic cisternae which are swollen and contain a finely flocculated material.

In addition to these changes affecting the entire chondriom, one can occasionally observe structural anomalies which are only found in some mitochondriae. In rare cases for instance we observed a swelling of the "matrix type" (THOENES, 1964) or a widening of the cristae together with an accumulation of a dense osmiophilic material in the intracristal space. In spite of all the various mitochondrial lesions visible in cells of the acinuscentral type, one can in only a few exceptional cases find indications for complete destruction of the mitochondriae. In these cases single mitochondriae or several together are surrounded by a single membrane and sometimes show considerable degenerative alterations.

Other Cytoplasmic Constituents

Some of the glycogen-free acinuscentral liver epithelia show a definite increase of microbodies without exhibiting marked structural anomalies of these characteristic cytoplasmic organelles: they mostly show a roughly granulated matrix and may contain a lamellar core. The outer membrane often seems to be continuous with the ER (see SVOBODA, 1966; ESSNER, 1967).

The number of "peribiliar dense bodies" (PALADE and SIEKEVITZ, 1956), which are nowadays grouped with the "lysosomes" (DE DUVE, 1963) by most authors, is also increased in some cells, especially when there is an increase in the population of Golgi-complexes, with which they are usually in close vicinity (Fig. 5). Their structure does not differ from that of the peribiliar dense bodies of a normal liver cell.

In some instances one also finds fat droplets of various sizes and very rarely one encounters pigment granules, siderosomes and larger "cytosegrosomes" (ERICSSON and TRUMP, 1964).

2. The Acinusperipheral Cytotoxic Pattern

a) Light Microscopy
Cytological Characteristics

The peripheral cytotoxic pattern, as opposed to the central pattern, is mainly characterized by an *enhanced storage of glycogen*. In large areas of the cytoplasm one finds particles that can be stained by PAS or Best's carmine (Figs. 21, 23 and 27). If the sections are treated with diastase, these particles can no longer be revealed; the cytoplasm now seems to be optically empty. A similar impression, evoking the pattern of plant cells, ensues if glycogen is eluted by aqueous fixatives or staining solutions (Figs. 2, 22 and 26). This may easily simulate the pattern of a cellular hydrops. The increase of the glycogen depots usually leads to a noticeable *increase in cytoplasmic volume*, which may reach several times its normal value (Fig. 26). The ergastoplasmic bodies are usually forced towards the periphery of the cell or to the vicinity of the nucleus by the accumulated glycogen (Figs. 3 and 22). Their number per unit volume is markedly reduced. As the size of the cytoplasm increases concomitantly it is difficult to decide whether this represents a true or a relative *reduction of the ergastoplasm*. Some of the cells, however, are completely devoid of ergastoplasmic bodies in the plane of section and one must, therefore, consider the possibility that at least in some of the storage cells a true reduction in the amount of ergastoplasm occurs. In the course of chronic intoxication a peculiar material, which is weakly acidophilic ("hyalinic"), appears in the cytoplasm of some of the glycogen-rich epithelia. This material cannot be detected with the light microscope in a normal liver cell. It usually looks like a fine or coarse network, occasionally it shows a whorllike structure (Figs. 24 and 25). The whorls correspond to the "RNP-free cytoplasmic whorls" ("RNP-freie zytoplasmatische Wirbelbildungen"), first observed by ALTMANN and OSTERLAND (1960) after chronic thioacetamide intoxication. These *acidophilic cytoplasmic structures*, the network as well as the whorls are equivalent to variously shaped membrane complexes of the smooth ER observed with the electron microscope (cf. PORTER and BRUNI, 1959; BRUNI, 1960; THOENES and BANNASCH, 1962; STEINER and BAGLIO, 1963; LAFONTAINE and ALLARD, 1964; ORTEGA, 1966). They will be described in detail in the next chapter. The farther these membrane complexes spread in the cytoplasm, the more the enhanced glycogen depots are usually reduced. In the later stages of the experiment one can often find diastase-resistant globules which stain positively with PAS; with Goldner's stain they appear bright red.

Histogenesis of the Alterations as a Function of Dose and Time

In all experimental series the enhanced glycogen storage and the concomitant changes in the cytoplasm develop preferentially in the 1st and 2nd zone of the "functional" liver acinus. Depending on the plane of section they are to be found in round, oval or irregularly shaped parenchymal zones which touch the periportal region on one side and the glycogen-free centroacinal zone (see pages 5 and 6) on the other (Figs. 2 and 26). The extent of the enhanced glycogen storage and the speed of its development depend on the concentration of the carcinogen and the time of its administration.

After continuous administration of a *12 mg%o NNM-solution (continuous experiment)* one can observe within 2 weeks after the beginning of the experiment a marked increase of the cytoplasmic glycogen deposits in the parenchymal arrays outlined above (Figs. 1b, 2 and 3a). With prolonged time of administration one not only observes a further accumulation of glycogen within a cell, but also an enlargement of the parenchymal storage zones which may eventually reach the central vein. A maximum of this development is observed after 70—120 days. The above mentioned acidophilic cytoplasmic structures (smooth membrane complexes) are clearly visible under the light microscope for the first time on the 56th day after the beginning of carcinogen administration. At that time one finds them in only a few cells, later they can be detected in many of them. One can best describe their localization by relating it to the zones of glycogen storage at one side and the glycogen-free centroacinal zones at the other side. If the plane of section is properly laid, one observes the following characteristic pattern (Fig. 28): the center of the storage zones is occupied by cells which show only a small increase in cytoplasmic volume; these cells are surrounded circularly or semicircularly by several layers of epithelia which are strikingly rich in glycogen and, therefore, noticeably enlarged. There follows another

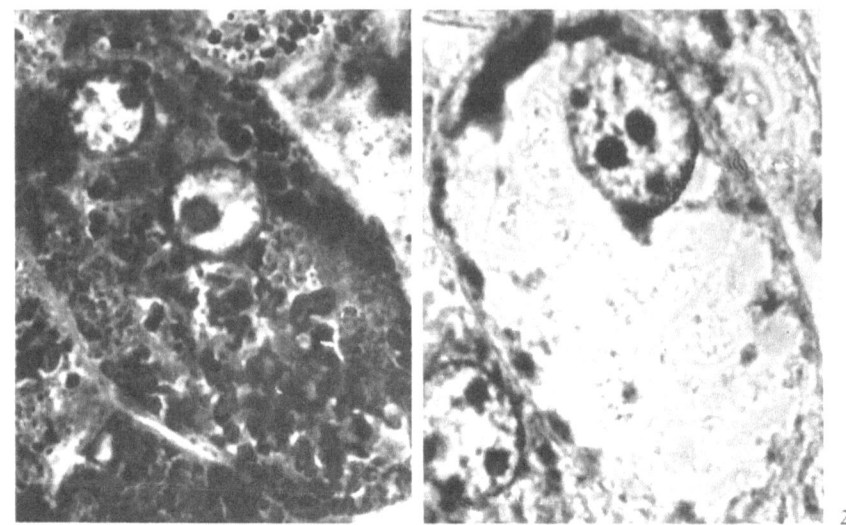

21   22

Fig. 21 and 22. Glycogen storage cells (acinusperipheral cytotoxic pattern) after chronic NNM-poisoning (12 mg% NNM-solution, 27 weeks). 21: Accumulation of coarse PAS-positive glycogen and considerable enlargement of the cytoplasm. Tri-PAS. 1400:1. 22: Elution of the stored glycogen during the preparation of the tissue leads to a plant cell-like picture. Few basophilic bodies in peripheral regions of the cytoplasm. Cresyl violet 1700:1

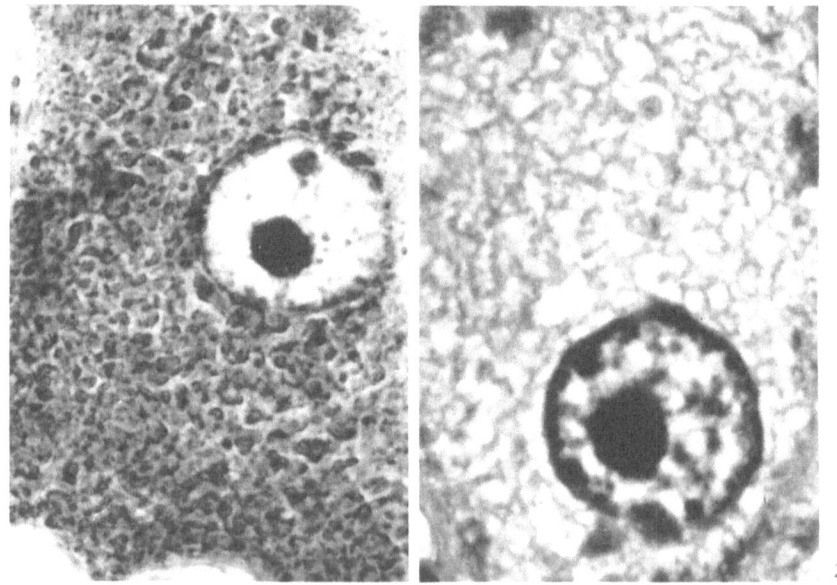

23   24

Fig. 23. Huge storage cell containing predominantly finely granulated glycogen. 12 mg%
NNM-solution, 19 weeks. Tri-PAS. 2000:1

Fig. 24. Large storage cell containing a loose acidophilic meshwork (hypertrophied agranular reticulum) troughout the cytoplasm. Finely granulated glycogen (see Fig. 23) eluted from the meshes of the network. "Dislocation" of the basophilic bodies towards the vicinity of the nucleus. 12 mg% NNM-solution, 19 weeks. HE. 1600:1

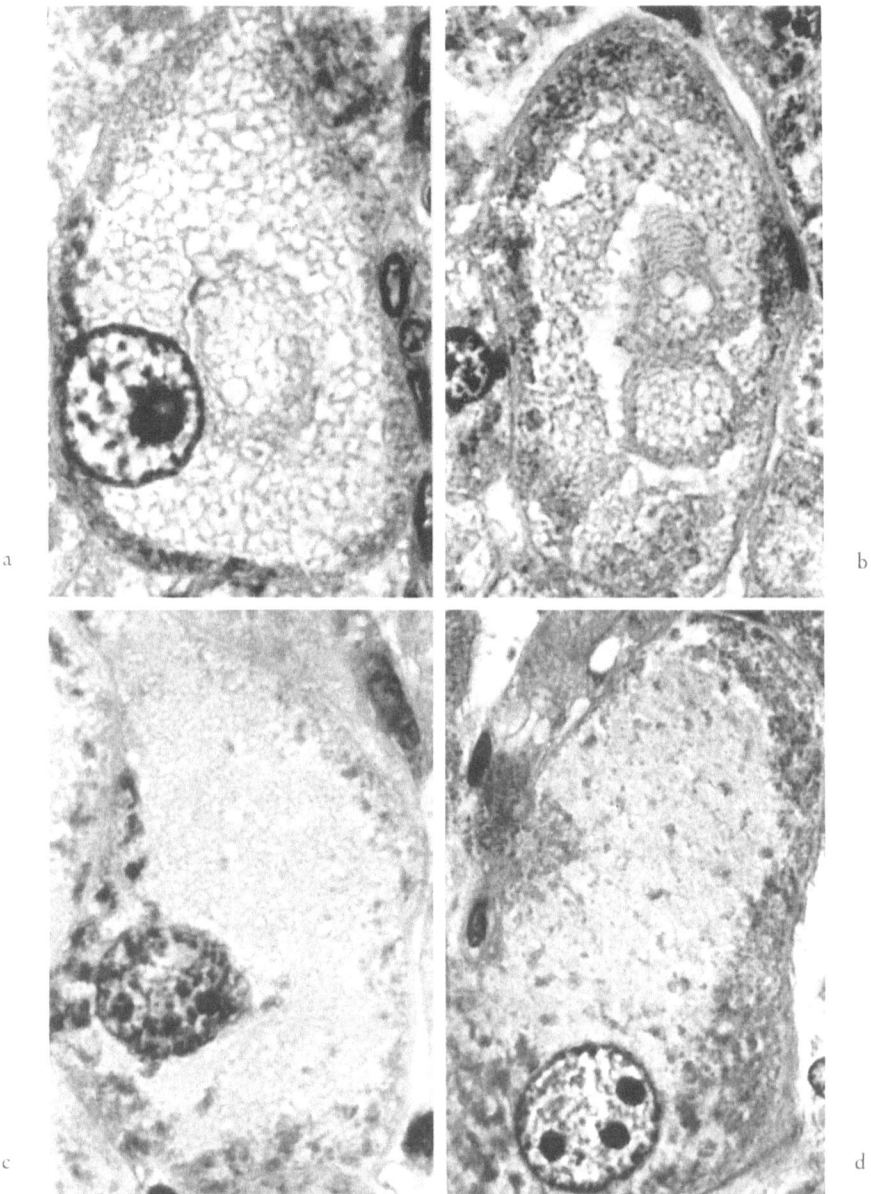

Fig. 25 a—d. Different types of hypertrophy of the agranular ER as seen under the light microscope (12 mg⁰/o NNM-solution, 19 and 27 weeks). a Loose acidophilic meshwork throughout the cytoplasm; in the center beginning formation of a whorl. Glycogen eluted from the meshes of the network. Very few basophilic bodies in the peripheral regions of the cytoplasm, near the cell nucleus. HE. 1100:1. b Somewhat denser meshwork with obvious double whorl in the center. Broad peripheral margin of ergastoplasm. HE. 1400:1. c. Dense acidophilic meshwork in a rather glycogen-poor cell. Few basophilic bodies in the peripheral regions of the cytoplasm. HE. 1300:1. d. Tightly packed acidophilic ("hyalinic") material within large areas of the cytoplasm. Broad peripheral margin of ergastoplasm. HE. 1100:1

layer of especially big cells which show a gradual increase of acidophilic material (smooth membrane complexes) and less glycogen towards the acinuscentral glycogen-free zone; the last layer is made up of a thin centroacinal parenchymal zone, almost

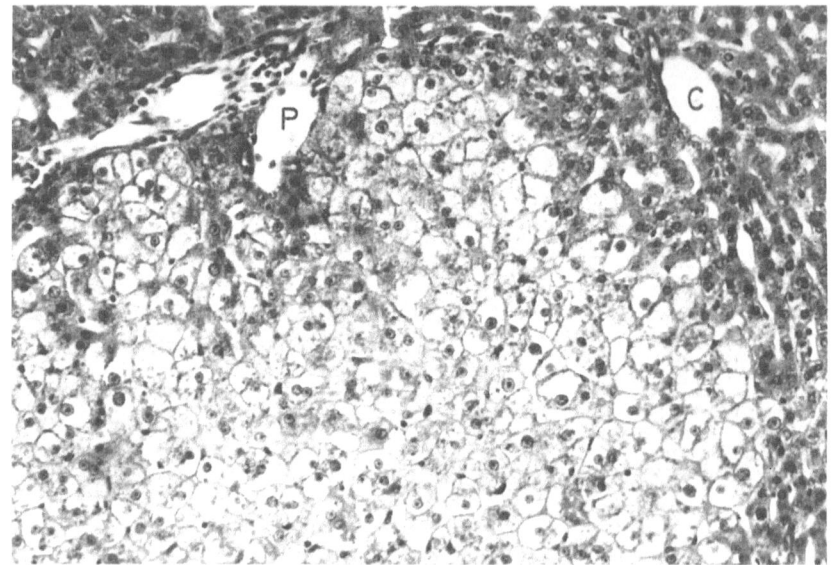

Fig. 26. Typical parenchymal focus of enhanced glycogen storage 15 weeks after removal of the carcinogen (20 mg% NNM-solution, 7 weeks). Considerably enlarged hepatocytes showing a clear cytoplasm after elution of glycogen during the preparation of the tissue. Periportal field (P). Central vein (C). Note that the storage focus does not reach the central vein (see text). HE. 200:1

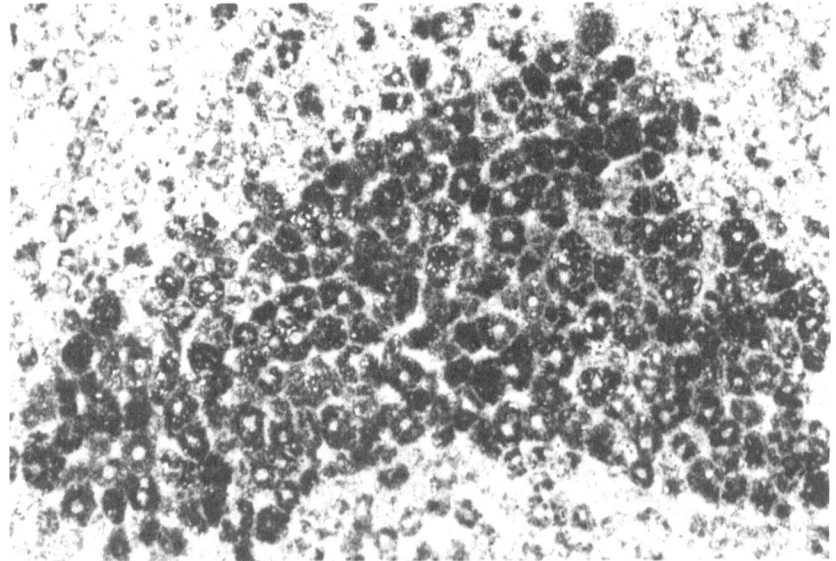

Fig. 27. Focus of liver parenchyma storing excessive glycogen 10 weeks after removal of the carcinogen (12 mg% NNM-solution, 12 weeks). Tri-PAS. 250:1

completely devoid of glycogen. The development of the various toxic cytoplasmic changes thus follows a definite course, whereby the smooth membrane complexes are preferentially formed in areas between those of maximal glycogen storage and almost

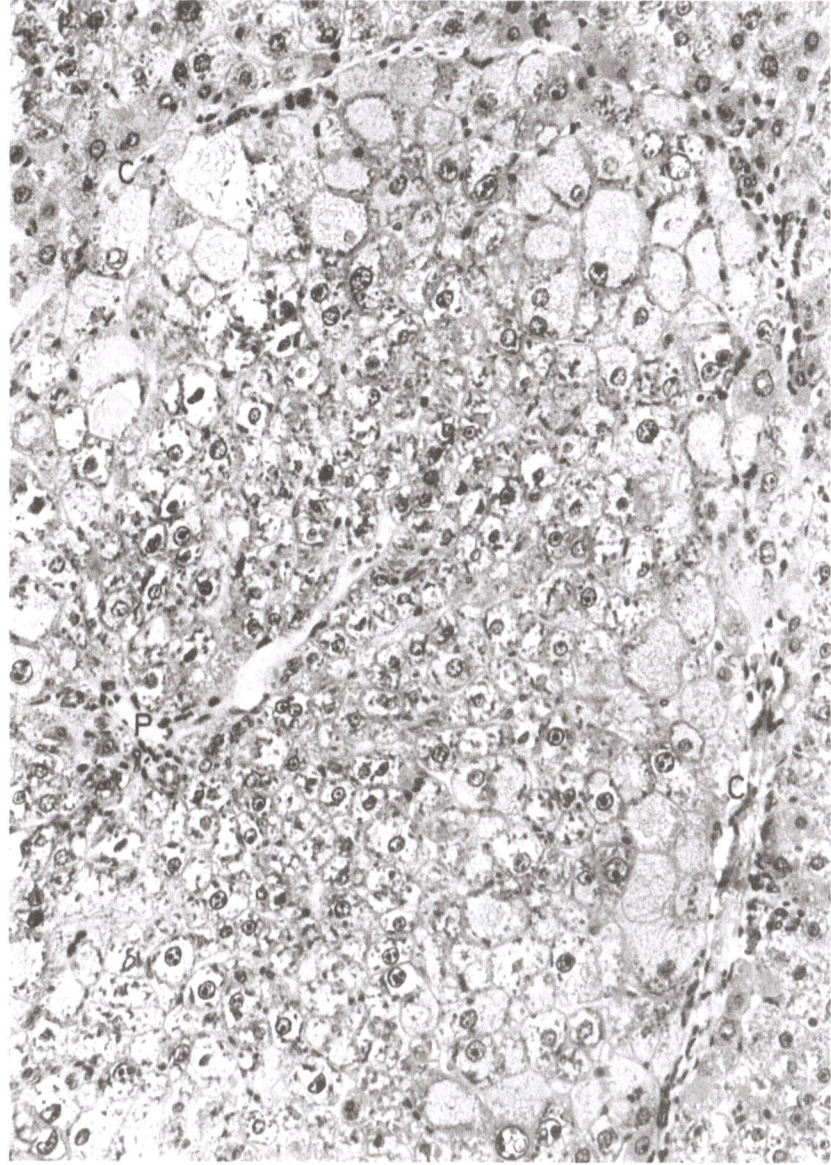

Fig. 28. Focus of liver parenchyma showing enhanced glycogen storage. In the immediate vicinity of the periportal field (P) epithelia contain nearly normal amounts of glycogen. Some layers of enlarged clear cells follow. They can be clearly distinguished at center left. At the periphery of the focus one observes a semicircular zone which reaches the central vein (C) and consists of especially big cells containing large amounts of acidophilic material (agranular membrane complexes of the ER). For detailed explanation see text. 12 mg%/o NNM-solution, 19 weeks. HE. 220:1

complete glycogen loss. Cells showing these peculiar cytoplasmic alterations thus occupy an intermediary position between those of the acinuscentral and the acinus-peripheral cytotoxic pattern. It should be mentioned, however, that one does find cells within the glycogen-free zone which do contain considerable amounts of acidophilic cytoplasmic structures, especially during later stages of the experiment. We presume that such cells move in from the periphery (see STÖCKER, 1966) in the course of the toxic treatment in order to replace cells that have undergone the above-mentioned necrosis. It is obvious that cells of this type may also undergo secondary changes typical of the acinuscentral toxic pattern, such as disaggregation of the basophilic bodies.

The "*stop-experiment*" shows very clearly that one can rightly differentiate between an acinuscentral and an acinusperipheral cytotoxic pattern. Whereas the cytoplasmic alterations of the central type usually disappear after removal of the carcinogen from the diet, those of the peripheral type persist for weeks and even months in many cells, including the increased smooth membrane complexes. Observations with the light microscope show very strikingly the persistence of the enhanced glycogen deposits. In the stop-experiment discussed here we were able to reveal the presence of multiple massive glycogen storage zones in the parenchyma of all animals sacrificed after the end of the carcinogen treatment (Fig. 17). In a further experiment under way we observed such zones as long as one year after the end of a nine-week administration of a 12 mg% NNM-solution.

If one administers a *6 mg% NNM-solution* instead of the 12 mg% solution, all cytoplasmic alterations characteristic of the peripheral cytotoxic pattern develop in basically the same way. They do, however, appear much later. Enhanced glycogen storage can be revealed for the first time only 14 weeks after the beginning of the experiment. At the same time, the first smooth membrane complexes are visible under the light microscope. Animals from experimental groups sacrificed later (after 22, 24 and 37 weeks) show parenchymal zones of enhanced glycogen storage which correspond in size to those observed after a 10—15 weeks uptake of a 12 mg% solution of NNM.

Administration of a *20 mg% NNM-solution* favours cytoplasmic changes of the acinuscentral type over the acinusperipheral type as long as the carcinogen is administered; thus, one can at first observe a marked depletion of glycogen in the liver parenchyma (Fig. 1 c). In small parenchymal zones, however, one can often witness a very pronounced glycogen storage, especially in peripheral areas of the lobule. After removal of the carcinogen from the diet (27 weeks after beginning the treatment with NNM) widespread parenchymal glycogen storage zones develop within 14 to 28 days, just as in the experiments described above. The storage zones are again observed with all animals sacrificed after the termination of the treatment. The results obtained in this series of experiments are of special interest in so far as they prove that under certain conditions cells of the acinuscentral cytotoxic pattern can give rise to cells of the acinusperipheral type.

b) Electron Microscopy

Glycogen and Endoplasmic Reticulum (ER)

The higher glycogen content in the cytoplasm seen with the light microscope is mirrored by a considerable *enlargement of the cytoplasmic glycogen zones*. Preparations fixed with OsO₄ and stained with uranyl acetate reveal such zones as light cloudy areas (Figs. 29, 33 and 34). They occupy large areas of the cytoplasm, in some cases at the center, in other cases at the periphery of the cell. After additional staining with lead hydroxide the glycogen appears in the form of small dense particles (Figs. 30, 45 and 47). They have the same size, shape and show the same arrangement as in normal liver cells (see DROCHMANS, 1960, 1962; THEMANN, 1964; REVEL, 1964). Their diameter ranges between 150 and 400 Å; they are embedded in the cytoplasmic matrix either singly (β-particles, DROCHMANS) or in the form of rosettes (α-particles, DROCHMANS). The diameter of the rosettes varies between 1000 and 2000 Å. If the sections are treated with uranyl acetate, one usually observes optically empty "holes" (STEINER and BAGLIO, 1963). In some cases, however, uranyl acetate leads to staining of the glycogen particles, which is similar to but much weaker than that obtained with lead hydroxide.

The enlarged glycogen zones are either completely free of cytoplasmic organelles (Fig. 29) or they contain individual cisternae of the ER, mitochondria, microbodies and sometimes even fat droplets. It is remarkable that frequently more or less important amounts of the stored *glycogen* and other cytoplasmic components are *separated from the rest of the cytoplasm by a single membrane* (Figs. 31 and 32). This leads to peculiar bodies containing a material which for the most part shows the same "staining reactions" as the glycogen zones. Lead hydroxide-stained preparations demonstrate the finely granulated monoparticulate appearance of glycogen within areas limited by the membrane (Fig. 32). Only rarely does one observe the formation of rosette-like aggregates. In addition to glycogen, one usually finds confined aggregates of a dense amorphous or granular material, irregularly or concentrically disposed membranous elements and cytoplasmic organelles, such as mitochondria. One can gradually follow the formation of such bodies via the segregation of certain cytoplasmic areas. The first step is the appearance of diffuse lines of demarcation in the cytoplasm. Later, distinct membranes are formed, which at first do not fully enclose the respective cytoplasmic area, whereby more or less wide connections to the surrounding cytoplasm are maintained (Fig. 31). Finally, even these connections are severed and the limiting membrane is completed (Fig. 32). This mode of segregation is typical of a frequently observed group of cytoplasmic bodies, called "*cytosegrosomes*" according to a suggestion by ERICSSON and TRUMP (1964). They are probably identical with the "*cytolysosomes*" (NOVIKOFF, 1961). One may rightly assume that the glycogen-containing cytosegrosomes described above actually do belong to the "lysosomes" (DE DUVE, 1963), as PHILLIPS et al. (1967) have recently reported the presence of the "key enzyme" of lysosomes, acid phosphatase, in such bodies in the liver of new-born rats. All possible transitions lead from those cytosegrosomes, containing mostly glycogen, to similar bodies which consist mainly of the above-mentioned dense amorphous or granular material and of unorganized membranes; in this case, glycogen is confined to smaller areas. All these cytoplasmic alterations are found almost exclusively during the later stages of the experiment, regardless of whether the carcinogen is still being administered or has been removed from the diet weeks before. We presume that they correspond to the PAS-positive spherical bodies which can be seen with the light microscope at about the same time in the cytoplasm of storage cells.

The ergastoplasm of storage cells is mainly confined to a small zone along the cell membrane or to the vicinity of the cell nucleus (Fig. 29 and 33). Its concentration

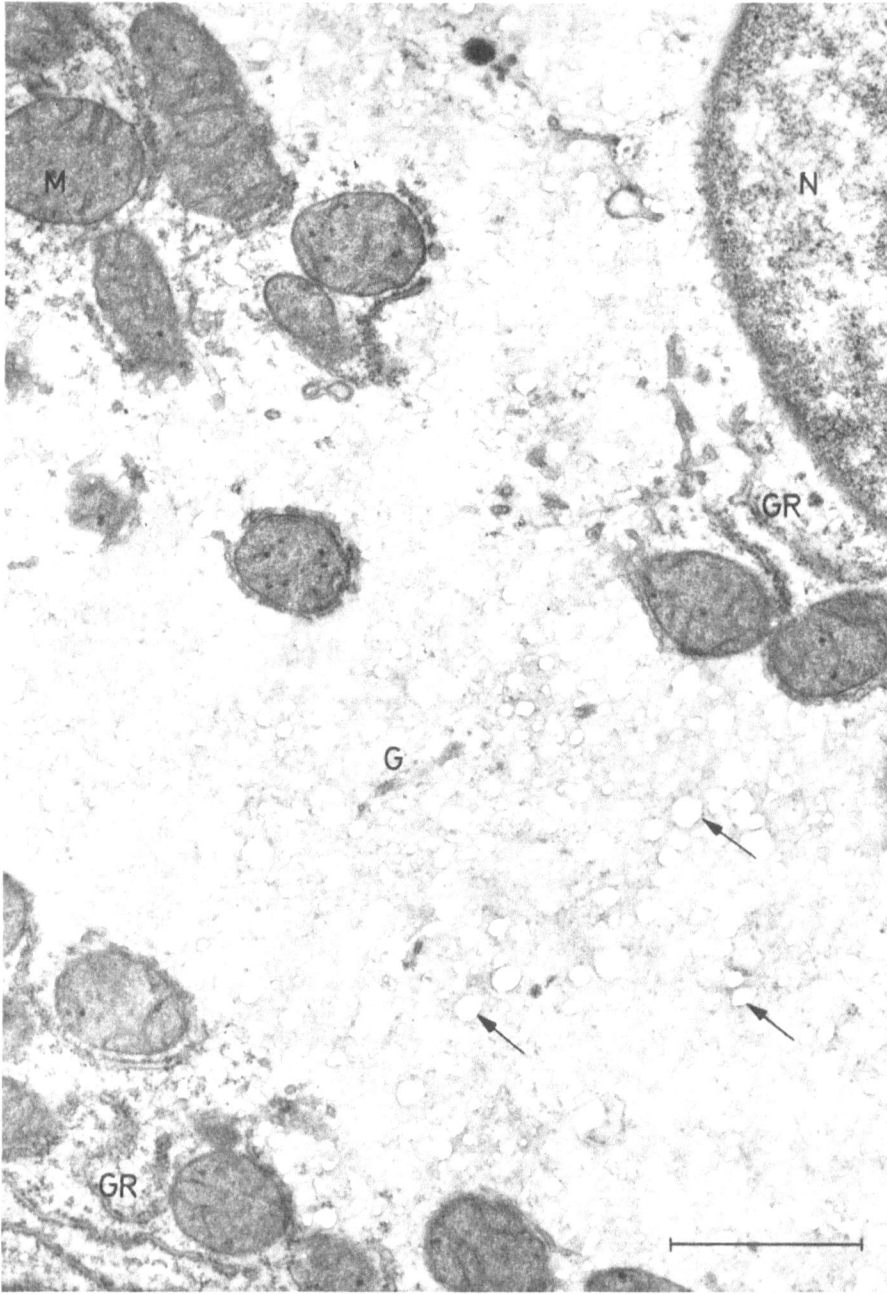

Fig. 29. Portion of a glycogen storage cell. Large cloudy glycogen zone (G) with multiple "glycogen holes" (↓). Granular reticulum (GR) and mitochondria (M) are forced towards peripheral regions of the cytoplasm or towards the vicinity of the nucleus (N). 12 mg⁰/₀ NNM-solution for 12 weeks, followed by plain water for 8 weeks. Uranyl acetate. 25 000:1

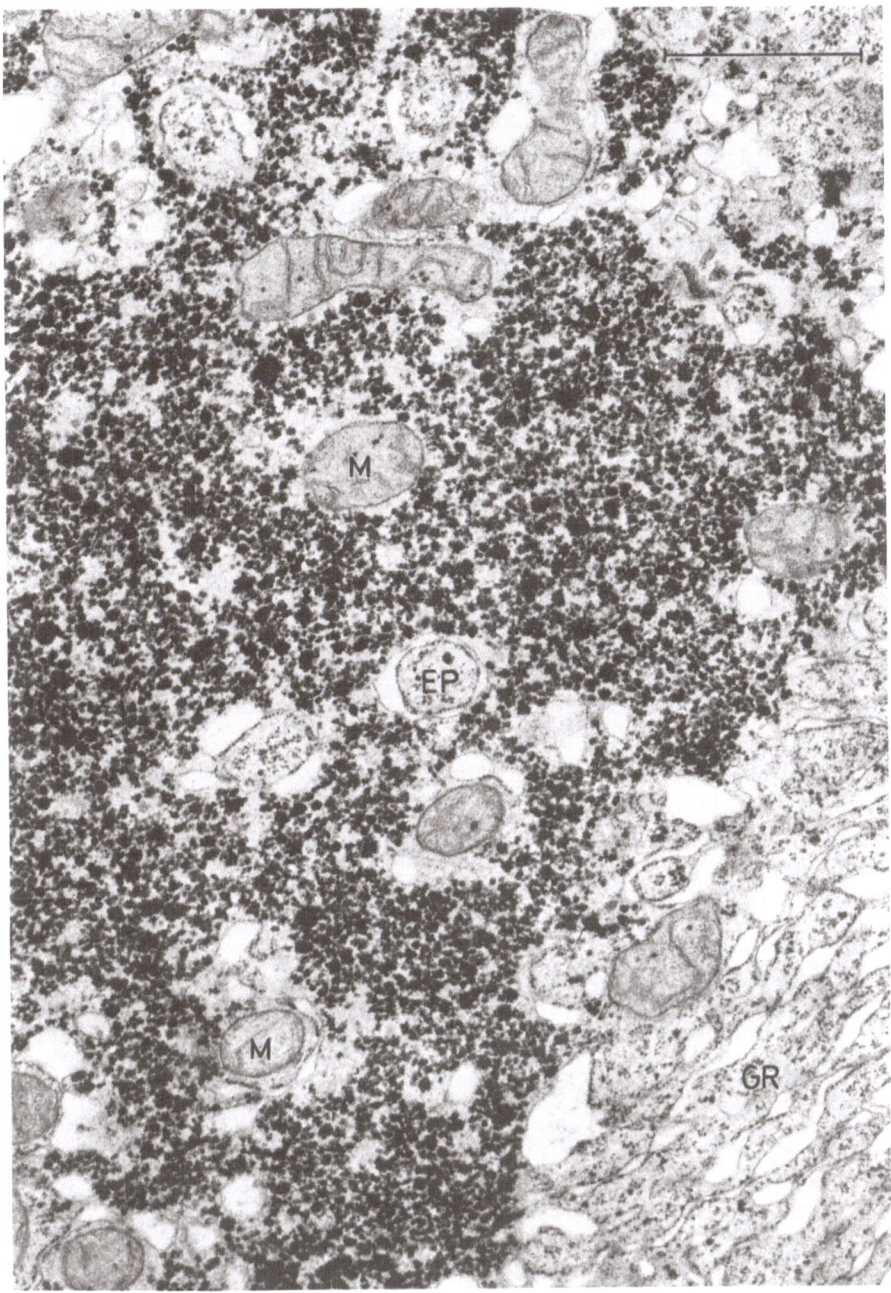

Fig. 30. Portion of a glycogen storage cell. Excessive accumulation of dense glycogen particles which form mostly rosettes (α-particles). Typical parallel arrangement of the granular cisternae (GR) which are sometimes slightly dilated. Two "ergastoplasm pockets" (EP) within the accumulated glycogen. Mitochondria (M) contain relatively few cristae. 20 mg%/o NNM-solution for 7 weeks, followed by plain water for 21 weeks. Lead hydroxide. 27 000:1

per unit volume of the cytoplasm is noticeably reduced. We have already pointed out the difficulties associated with evaluating the meaning of the reduction of the ergasto-plasm which may be relative or absolute. It is of special interest that the ergastoplasmic ultrastructure of storage cells is not at all or only very slightly altered. The parallel arrangement of the rough cisternae is retained in most cells (Figs. 30, 33 and 36). Individual cisternae seem to be unaffected, but may sometimes be partly dilated. Some

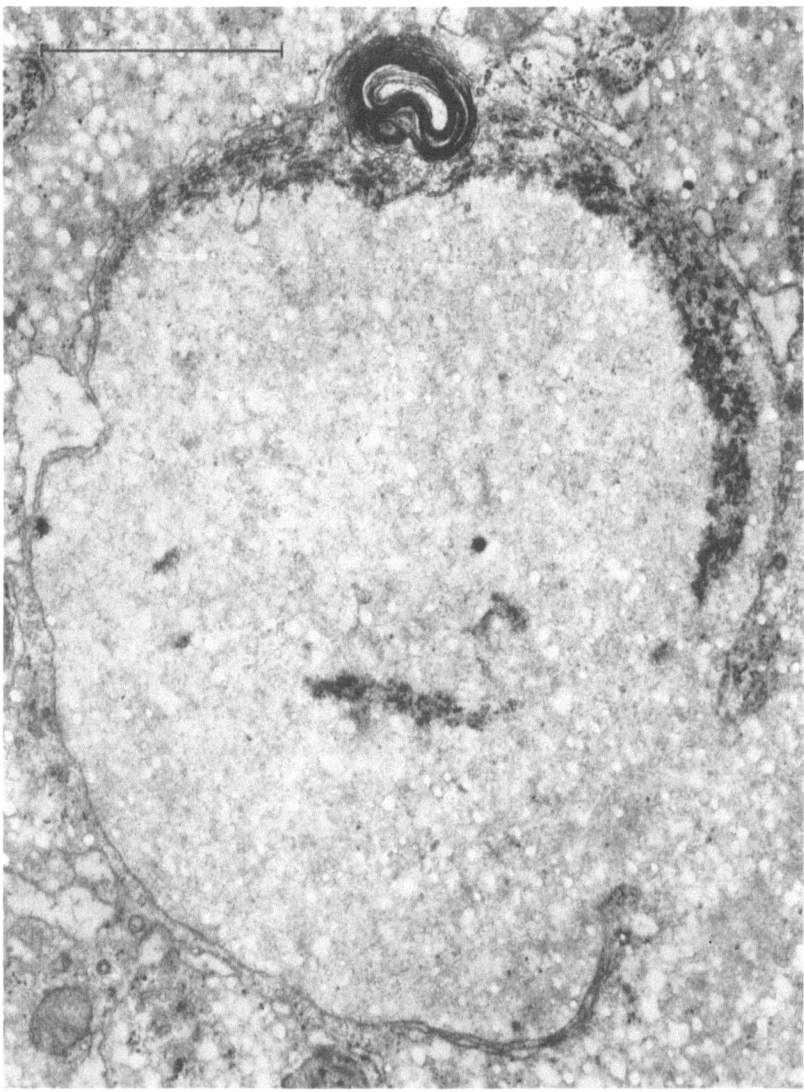

Fig. 31. Portion of a glycogen storage cell. Partial segregation of a glycogen-containing cytoplasmic area by a single membrane. A broad connection with the surrounding cytoplasm remains (bottom right). Concentrically arranged membranes within the segregated area (upper center). In addition some granular osmiophilic material. 20 mg⁰/o NNM-solution for 7 weeks, followed by plain water for 21 weeks. Uranyl acetate. 33 500:1

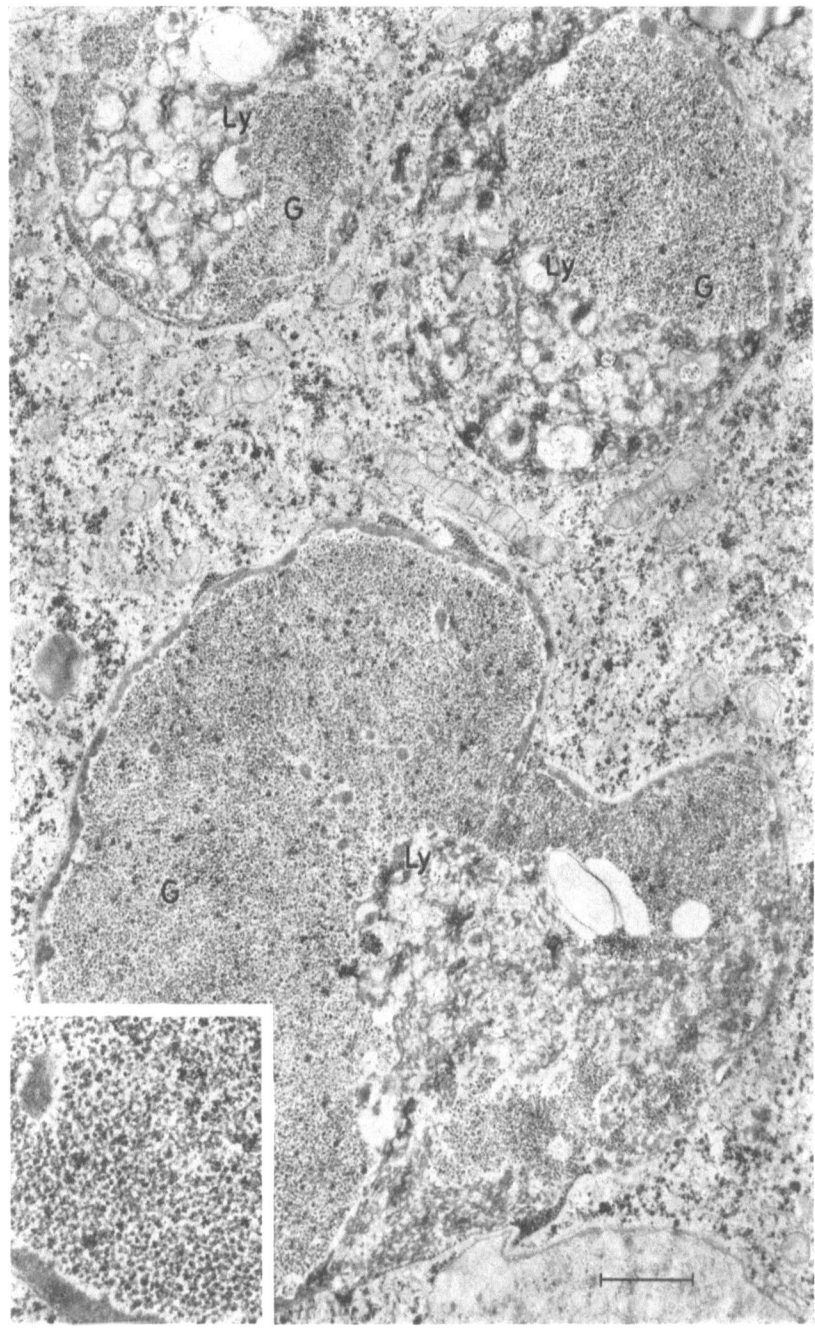

Fig. 32. Portion of a glycogen storage cell. Three large cytolysosomes (Ly) surrounded by a single membrane and containing masses of finely granulated monoparticulate glycogen (G) together with degraded cytoplasmic organelles and amorphous osmiophilic material. Abundant glycogen, mostly forming rosettes, in the surrounding cytoplasm. Mitochondria (M). 20 mg⁰/₀ NNM-solution for 7 weeks, followed by plain water for 21 weeks. Lead hydroxide. 13 000:1.
      Inset: Portion of the lower cytolysosome: finely granulated glycogen. 31 500:1

ergastoplasm aggregates show slight structural changes insofar as their cisternae have a more wavy or meandering path. In rare cases one may observe concentric lamellar formations of the rough cisternae (rough or ergastoplasmic "Nebenkern", HAGUENAU, 1958) (Fig. 40). These membranous formations are never observed in normal liver cells, but have been described earlier for hepatoma cells (FAWCETT and WILSON, 1955; ROUILLER, 1957; DRIESSENS et al., 1959). The disorganization of the ergastoplasm, typical of the acinuscentral cytotoxic pattern, is only occasionally observed with the strictly acinusperipheral type and only in the form of a displacement of otherwise unchanged rough cisternae. Thus, there are two main features of the ergastoplasmic changes in storage cells: 1. a shift or *dislocation of ergastoplasmic bodies* which may simply be caused by the mechanical stress exerted by the accumulated glycogen and 2. at least a *relative reduction of the ergastoplasm*.

The smooth ER of many storage cells is only weakly developed, just as in the case of normal liver cells. The typical network (FAWCETT, 1955), made up of tubular elements, is disposed along the borders of the glycogen zones, in the vicinity of the ergastoplasm or the cellular membrane. Where the network is in contact with the ergastoplasm zones, the smooth tubular elements often communicate with the flat, rough cisternae. In the holes of the network glycogen particles are inserted which usually are in intimate contact with the membranes. Upon sectioning the smooth ER often looks like small vesicles, which are irregularly distributed over the glycogen zones. The vesicles often contain dense osmiophilic particles. When the results of optical microscopy were described above, we mentioned already the development of large smooth membrane complexes in a number of storage cells during chronic intoxication. At the same time the amount of accumulated glycogen usually decreases, the decrease being the more pronounced the farther the smooth ER spreads. All stages ranging from cells with enhanced glycogen storage, but little AR, to those rich in AR, but relatively poor in glycogen, can be observed with the electron microscope.

As long as the increase in membranous material is relatively small, the cells contain large amounts of glycogen, far more than normal liver cells. Under these conditions the smooth ER is arranged in defined arrays of various shapes. In most cases it extends from the periphery of the cell to the glycogen zones, without a distinct separation from the accumulated glycogen (Fig. 34). It retains the characteristic structure of a network, the only difference between it and the AR of a normal liver cell being the fact that it covers a bigger area. We may, therefore, speak of a *"hypertrophy" of the smooth ER*, as suggested by PORTER and BRUNI (1959). In other cases the membrane complexes develop mostly within the glycogen deposits (Figs. 33, 36 and 37), where one can again observe the typical network formations. They may, however, be partly surrounded by one or two long and smooth cisternae, thus forming a *"coiled network"* (*"Netzwerkknäuel"*, Figs. 36, 37), a phenomenon similar to the one described earlier for chronic intoxication with thioacetamide (THOENES and BANNASCH, 1962). The agranular flat cisternae are either continuous with tubular elements inside the coil or with rough cisternae on the outside. Sometimes one finds long smooth profiles inside the network which also continue in tubular elements (Fig. 36). In very rare cases the tubuli are swollen into vesicles. The size of the "coiled network" varies within wide limits, the diameter ranging between 0,5 and 12 μ. All formations of the ER described so far are closely associated with glycogen particles (α-or β-particles) which are inserted into the network or lined up along the elongated profiles (Fig. 35).

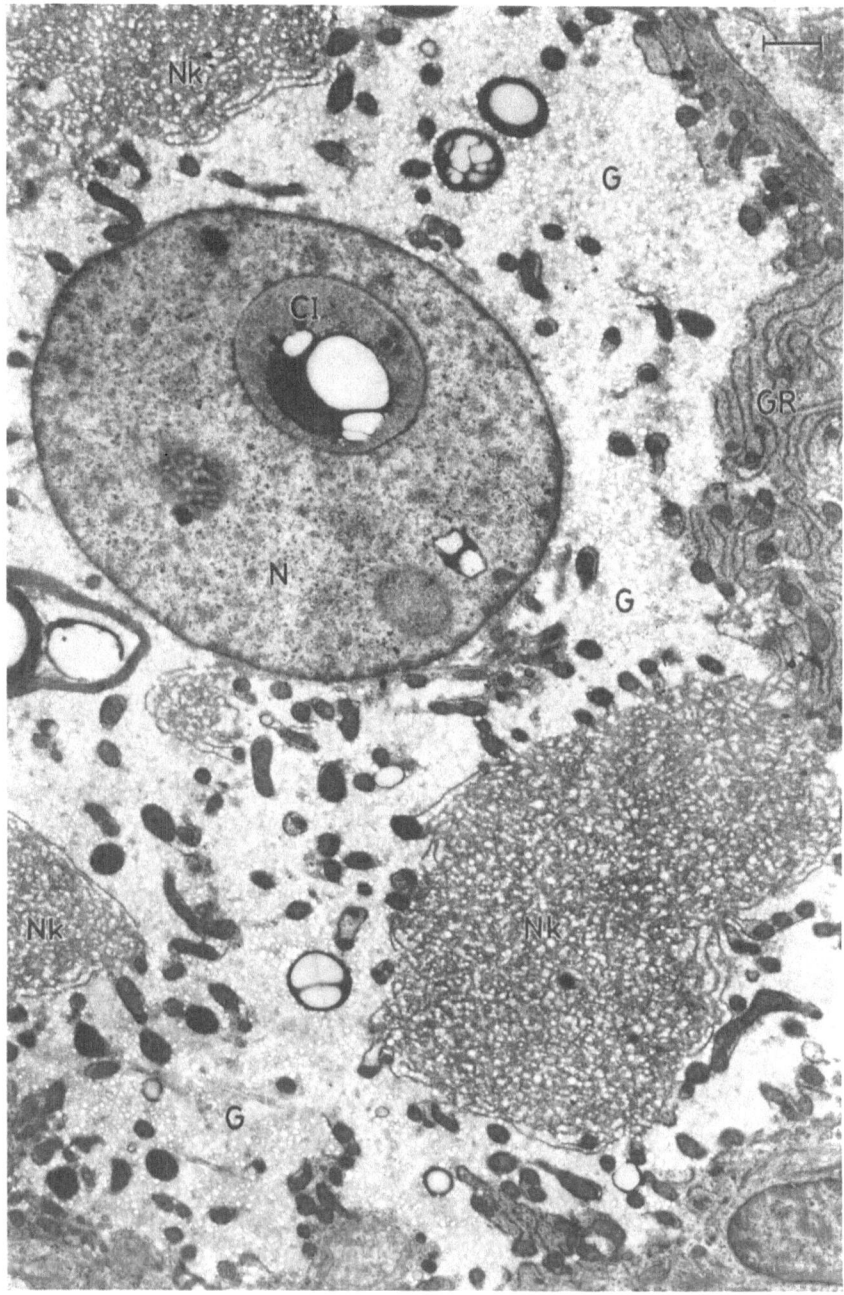

Fig. 33. Big glycogen storage cell. Large cloudy glycogen zones (G). Protruding into the glycogen zones several "Netzwerkknäuel" (Nk) ("coiled network") of the agranular reticulum (AR). Aggregate of granular cisternae (GR) which show a wavy path (middle right). Nucleus (N). Cytoplasmic inclusion (CI). 12 mg% NNM-solution for 12 weeks, followed by plain water for 8 weeks. Uranyl acetate. 16 000:1

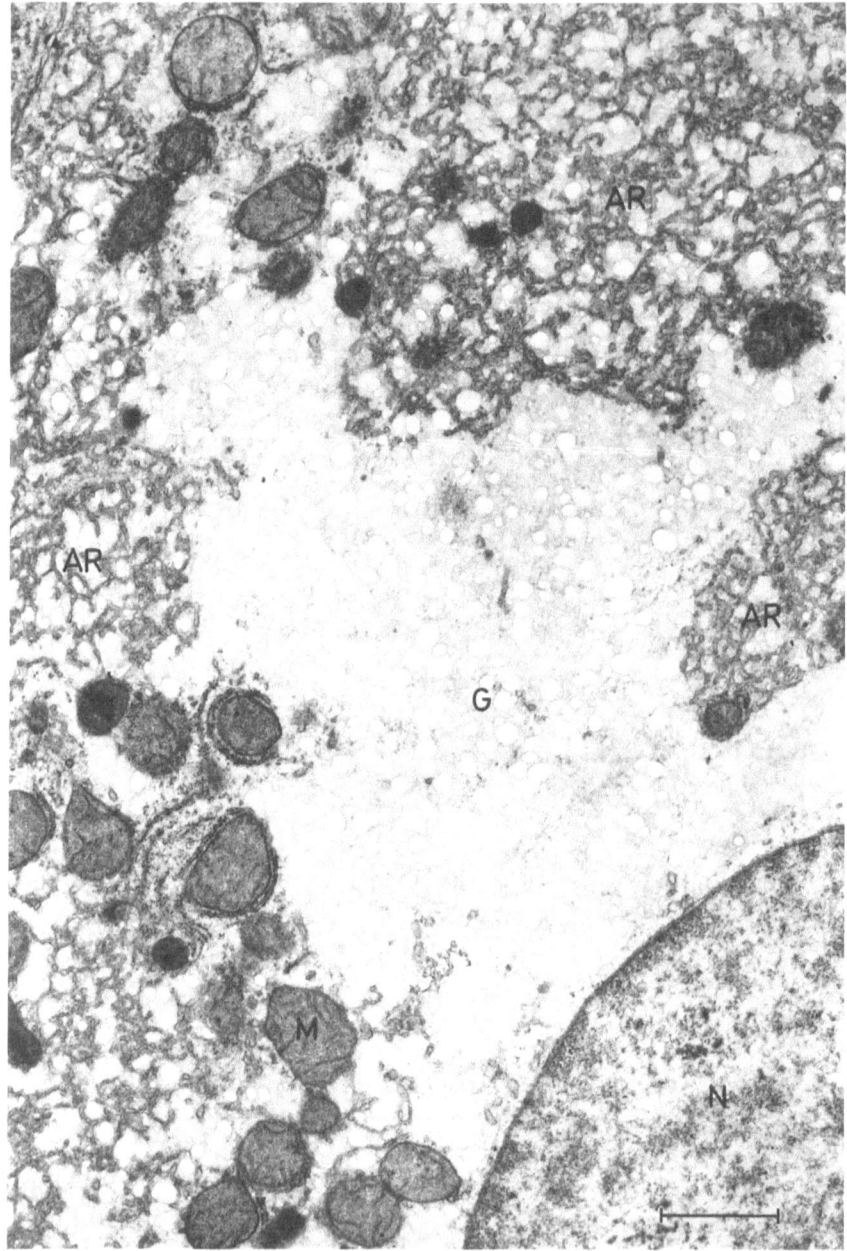

Fig. 34. Portion of a glycogen storage cell. Large cloudy glycogen zone (G). From the periphery of the cell a network consisting of tubular agranular elements (AR) extends into the glycogen zone without a distinct separation from the accumulated glycogen. Mitochondria (M). Nucleus (N). 12 mg⁰/o NNM-solution for 12 weeks, followed by plain water for 8 weeks. Uranyl acetate. 16 000:1

In addition to the "coiled network", we also observe within the glycogen zones of storage cells peculiar *"lamellar cistern complexes" of the agranular ER*, which are concentrically layered or follow a sinuous path (Figs. 38, 39 and 41). The smooth lamellae are often connected with tubular elements of the typical smooth network or with rough cisternae. Concentrically layered smooth membrane complexes have first

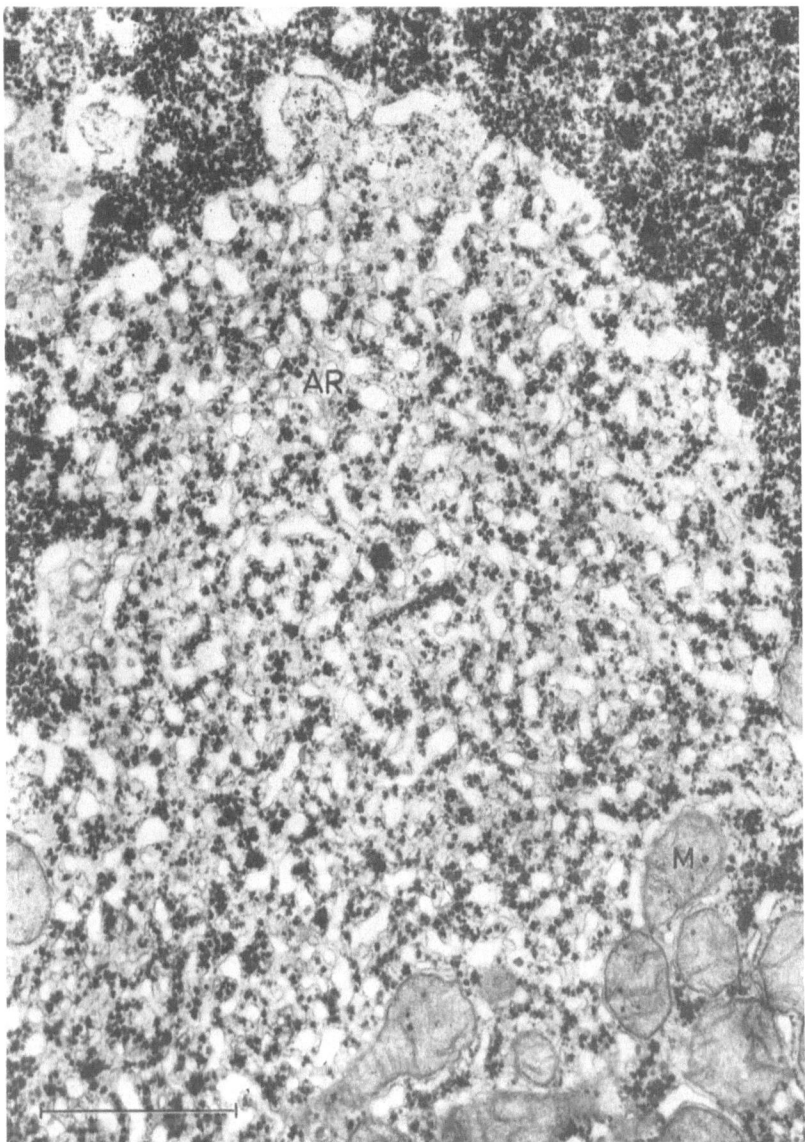

Fig. 35. Portion of a glycogen storage cell. Confined complex of tubular elements of the agranular reticulum (AR). In the vicinity and in the meshes of the network many glycogen particles mostly aggregated to rosettes. Mitochondria (M). 20 mg% NNM-solution for 7 weeks, followed by plain water for 21 weeks. Lead hydroxide. 26 000:1

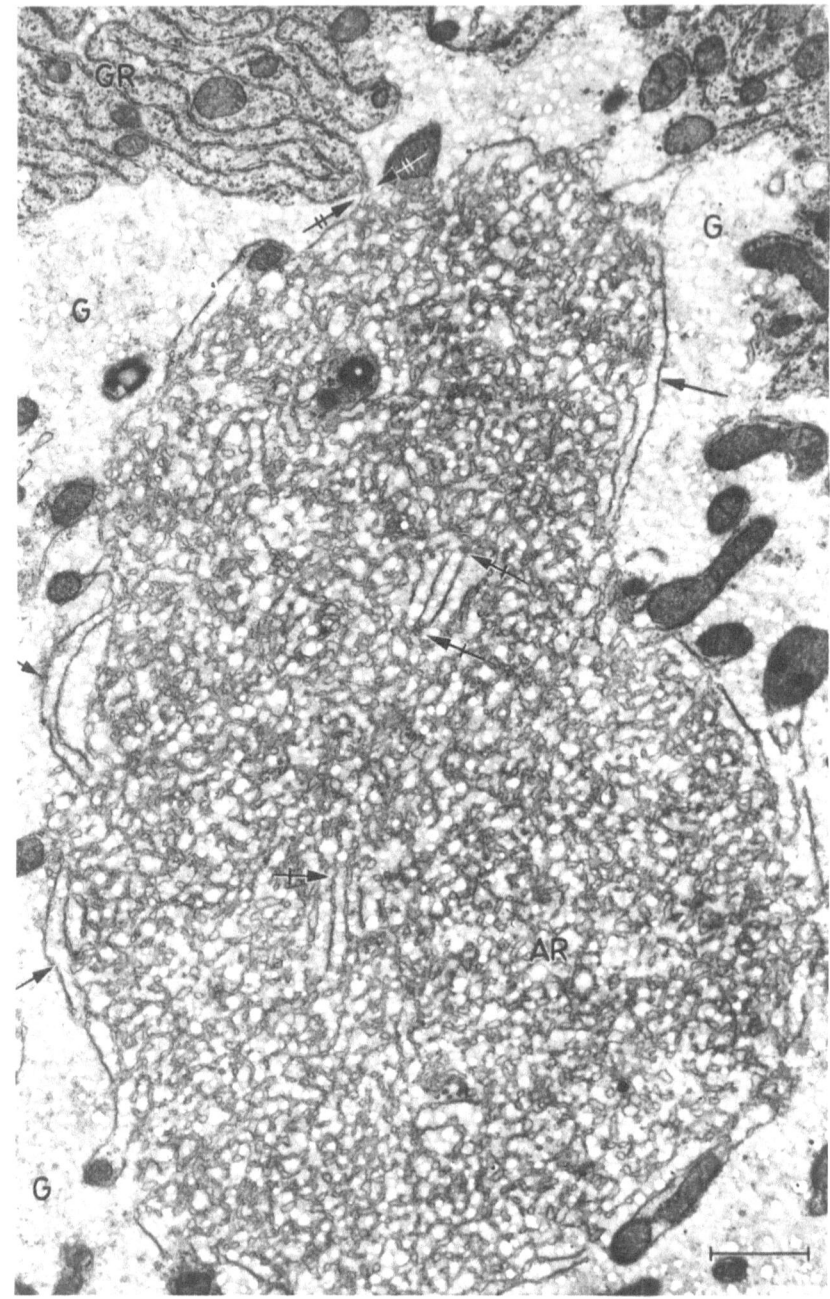

Fig. 36. Portion of a glycogen storage cell. "Netzwerkknäuel" ("coiled network") of the agranular reticulum (AR): the network consisting mainly of tubular elements is surrounded by some lamellar cisternae (↓) which are continuous with the tubular structures. Inside the membrane complex some tubular elements are also continuous with lamellae (†). Multiple "glycogen holes" are seen in the meshes of the network and in the surrounding glycogen zones (G). At the upper margin one can see a typical aggregate of granular cisternae (GR). At one point there is a direct connection (‡) between granular and agranular membranes. 12 mg⁰/o NNM-solution for 12 weeks, followed by plain water for 8 weeks. Uranyl acetate. 13 500:1

been described for the rat liver after low protein diet (CLEMENTI, 1960) and after exposure to the hepatocarcinogenic dimethylnitrosamine (BENEDETTI and EMMELOT, 1961). Later they were described under various names after treatment with different carcinogens and even non-carcinogenic liver toxins (SALOMON, 1962; THOENES and BAN-

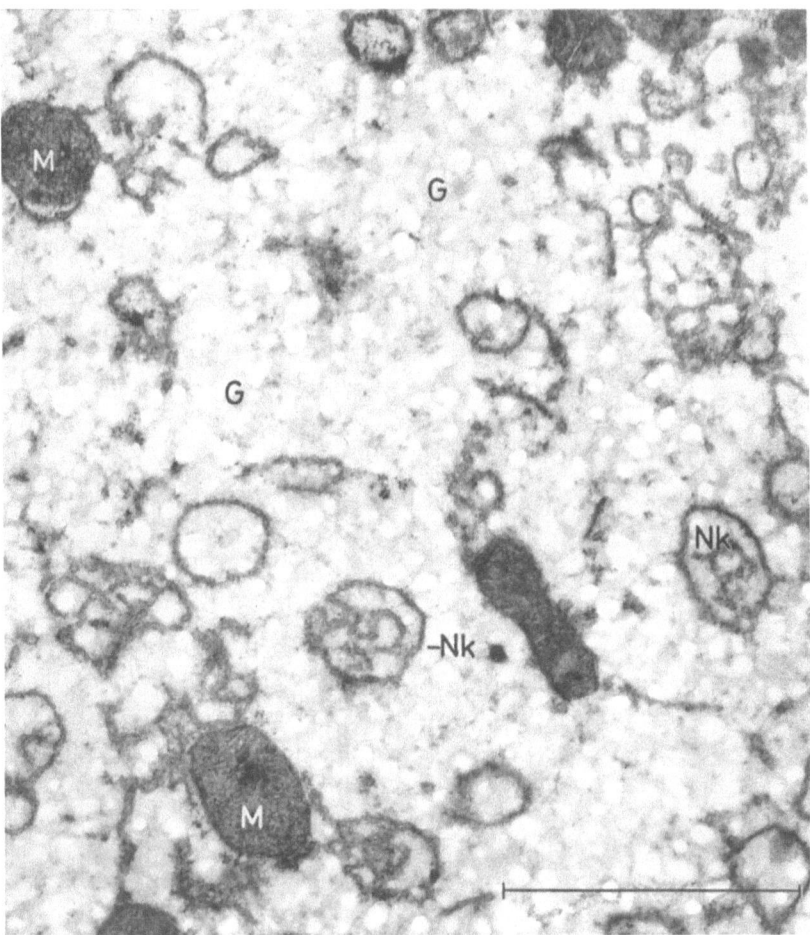

Fig. 37. Portion of a glycogen storage cell. Small "Netzwerkknäuel" (NK) ("coiled network") of the agranular reticulum within a cytoplasmic zone showing many "glycogen holes". Mitochondria (M). 12 mg⁰/o NNM-solution, 16 weeks. Uranyl acetate. 40 000:1

NASCH, 1962; TIMME and FOWLE, 1963; STEINER and BAGLIO, 1963; HERDSON et al., 1964; STEINER et al., 1964; STENGER, 1966; ORTEGA, 1966; and others). Our earlier suggestion that these concentrically layered membrane complexes (smooth "Neben-kerne") are part of a hypertrophy of the smooth ER (THOENES and BANNASCH, 1962) has in the meantime been adopted by most workers in the field (STEINER et al., 1964; HERDSON et al., 1964 a; ORTEGA, 1966; STENGER, 1966). The same explanation seems to apply to morphologically similar phenomena in the tubular epithelia of X-irradiated

3*

kidneys (LEAK and ROSEN, 1966) and in human fibromyxosarcomas (LEAK et al., 1967). STEINER and coworkers have already pointed out that the smooth cistern complexes of the liver epithelia appear in two different forms: they may either be

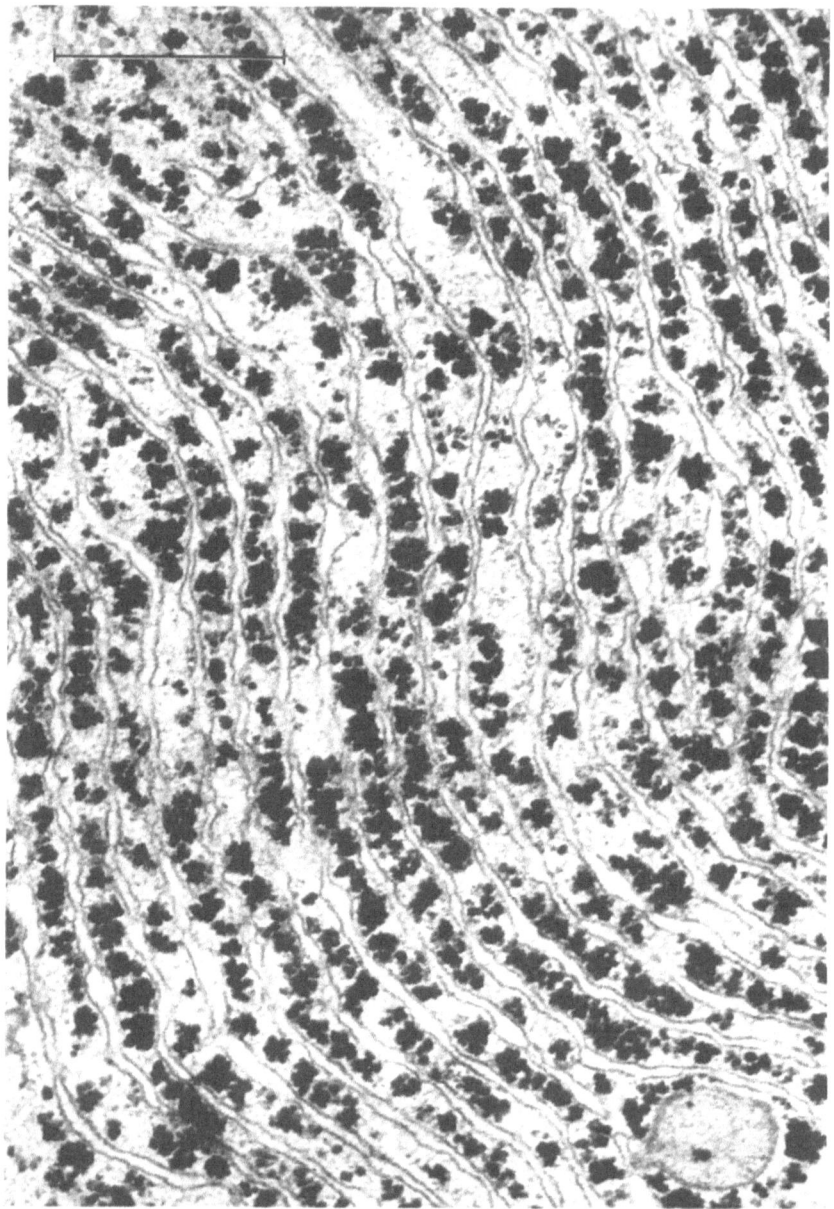

Fig. 38. Portion of a glycogen storage cell. "Lamellar cistern complex" of the agranular reticulum arranged in waves. In between the parallelly stacked smooth cisternae many glycogen rosettes (α-particles) or single glycogen particles (β-particles). 20 mg% NNM-solution for 7 weeks, followed by plain water for 21 weeks. Lead hydroxide. 31 500:1

associated with glycogen particles or may be completely free of them. In the first instance the cisternae show a rather loose arrangement, alternating with one or several layers of glycogen particles (α- or β-particles) (Figs. 38 and 39). In the second they

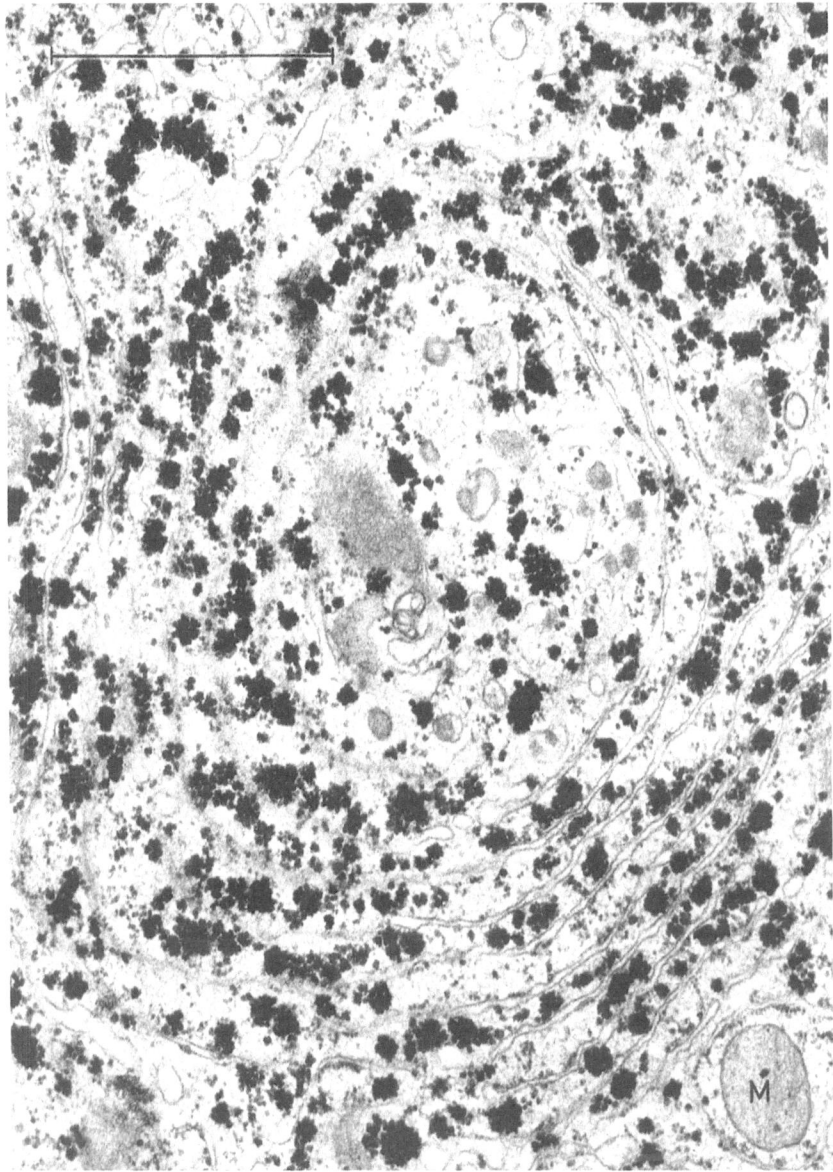

Fig. 39. Portion of a glycogen storage cell. Concentrically arranged "lamellar cistern complex" of the ER. In between the predominantly agranular membranes dense glycogen rosettes (α-particles). Sometimes glycogen particles are absent, while ribosomes are attached to the membranes. Mitochondria (M). 20 mg%/o NNM-solution for 7 weeks, followed by plain water for 21 weeks. Lead hydroxide. 38 000:1

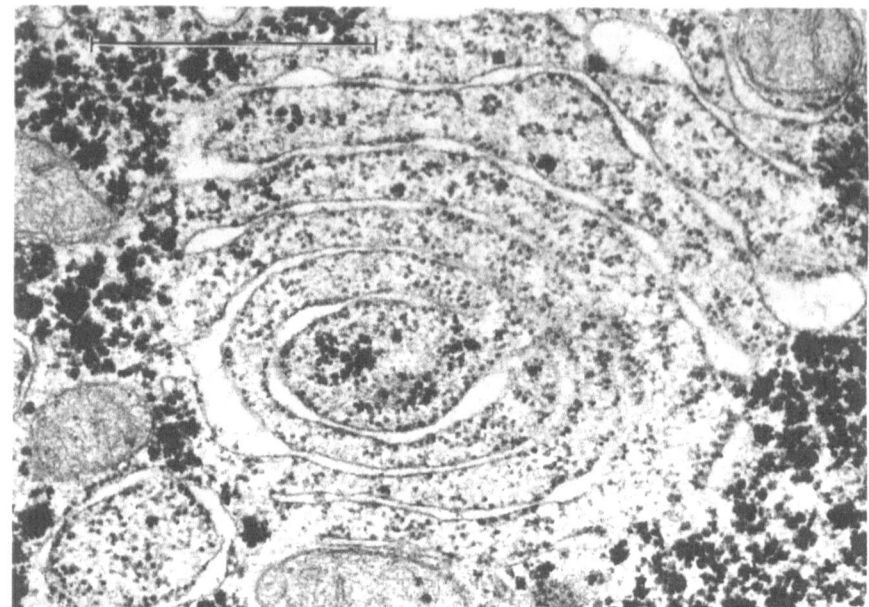

Fig. 40. Portion of a glycogen storage cell. Concentrically arranged "lamellar cistern complex" of the granular reticulum (so-called ergastoplasmic or rough "Nebenkern"). 20 mg⁰/₀ NNM-solution for 7 weeks, followed by plain water for 21 weeks. Lead hydroxide. 39 500:1

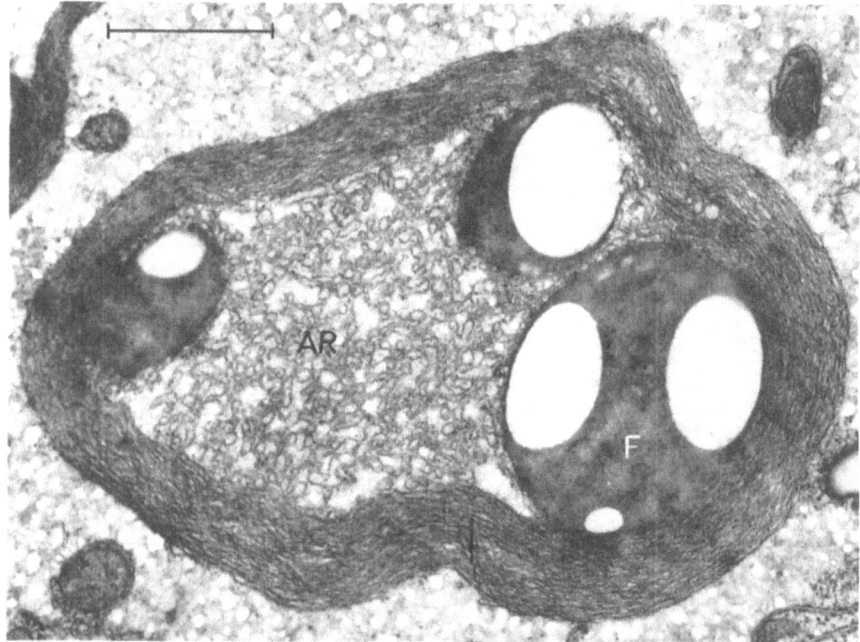

Fig. 41. Portion of a glycogen storage cell. Concentrically arranged "lamellar cistern complex" of the agranular reticulum (so-called smooth "Nebenkern"). The densely packed smooth lamellae surround three fat droplets (F) and a typical smooth network complex consisting of tubular elements which at many places are continuous with the surrounding lamellae. 12 mg⁰/₀ NNM-solution for 12 weeks, followed by plain water for 8 weeks. Uranyl acetate. 22 000:1

are densely packed (Fig. 41). We doubt whether or not it is useful to differentiate the membrane complexes which are associated with glycogen particles as "glycogen bodies" (STEINER et al., 1964) from the glycogen-free formations which are often referred to as "fingerprints". They seem to be only different expressions of the same basic phenomenon. It is further noteworthy that the concentrically layered cistern complexes usually enclose smaller areas of network (Fig. 41), fat droplets (Fig. 41) or mitochondria.

The increase in the amount of smooth ER can proceed to such an extent that it covers wide areas of the cytoplasm (Figs. 42 and 43) (see also PORTER and BRUNI, 1959; STEINER and BAGLIO, 1963; LAFONTAINE and ALLARD, 1964). Glycogen particles become scarce and can be found only in the meshes of the smooth ER or its immediate vicinity; some cells are completely devoid of them. In comparatively glycogen-poor cells the smooth ER also consists mainly of tubular network formations which often communicate with rough cisternae. The lumen of the tubular elements may either be strongly restricted or noticeably dilated. Upon sectioning this leads either to an almost inextricable membranous concentration or to large areas of approximately circular, oval or elongated membrane profiles. The tubular formations may again be continuous with lamellar cistern complexes which may reach a considerable size. The rough ER of such modified epithelia shows alterations in size and disposition similar to those observed in cells with enhanced glycogen storage: it seems to be reduced and is mostly confined to the periphery of the cell or to the vicinity of the cell nucleus. It may show the characteristic arrangement of the basophilic bodies, again stressing the similarity to the acinusperipheral cytotoxic pattern. In some cells, however, the rough ER is disorganized as in the glycogen-free or glycogen-poor epithelia of the centroacinal toxic pattern. This further indicates an intermediary position of AR-rich but glycogen-poor cells between the acinuscentral and the acinusperipheral toxic pattern, as previously suggested by light microscopy.

In the case of the alterations described above the smooth ER could be distinguished rather easily from the rough ER, in spite of frequently observed transitions from smooth- to rough-surfaced membranes. This need not always be the case. During the later stages of the experiment many glycogen-rich cells develop *peculiar combinations of smooth and rough ER*. In this case one has to differentiate between individual and more or less isolated cisternae and complex membrane formations. The *"combined cisternae"* always lie within the glycogen zones (Figs. 44, 45 and 47). Upon sectioning they look either like elongated meandering profiles or like "open" U-shaped circular or oval ones. The long profiles can be followed continuously over long distances. In rare instances, however, one may observe short interruptions, obviously pores. One can often notice that the profiles consist of two individual membranes. From these observations one may conclude that such long profiles are sections through broad flat cisternae with occasional pores (see also GIRBARDT, 1966). Their lumen can in some cells hardly be seen (Fig. 44), in others it is more or less dilated (Fig. 47). In the latter case it seems to be optically empty or else filled with a finely flocculated material; on rare occasions one observes individual dense osmiophilic particles. The "combined cisternae" are continuous with smooth network or with rough cisternae at the border of the glycogen zones. In cells especially rich in glycogen, the long profiles are mostly free of ribosomes and their smooth parts are invariably in close contact with glycogen particles (Figs. 44 and 45). At some points, however, the profiles are lined with ribosomes, usually unilaterally. The ribosomes are located preferentially in areas where the cisternae show inpocketing (Figs. 45, 46 and 47). In addition to the membrane-bound ribosomes one usually encounters in the *"ergastoplasm pockets"* free ribosomes, mitochondria or microbodies. Glycogen particles are rare or completely

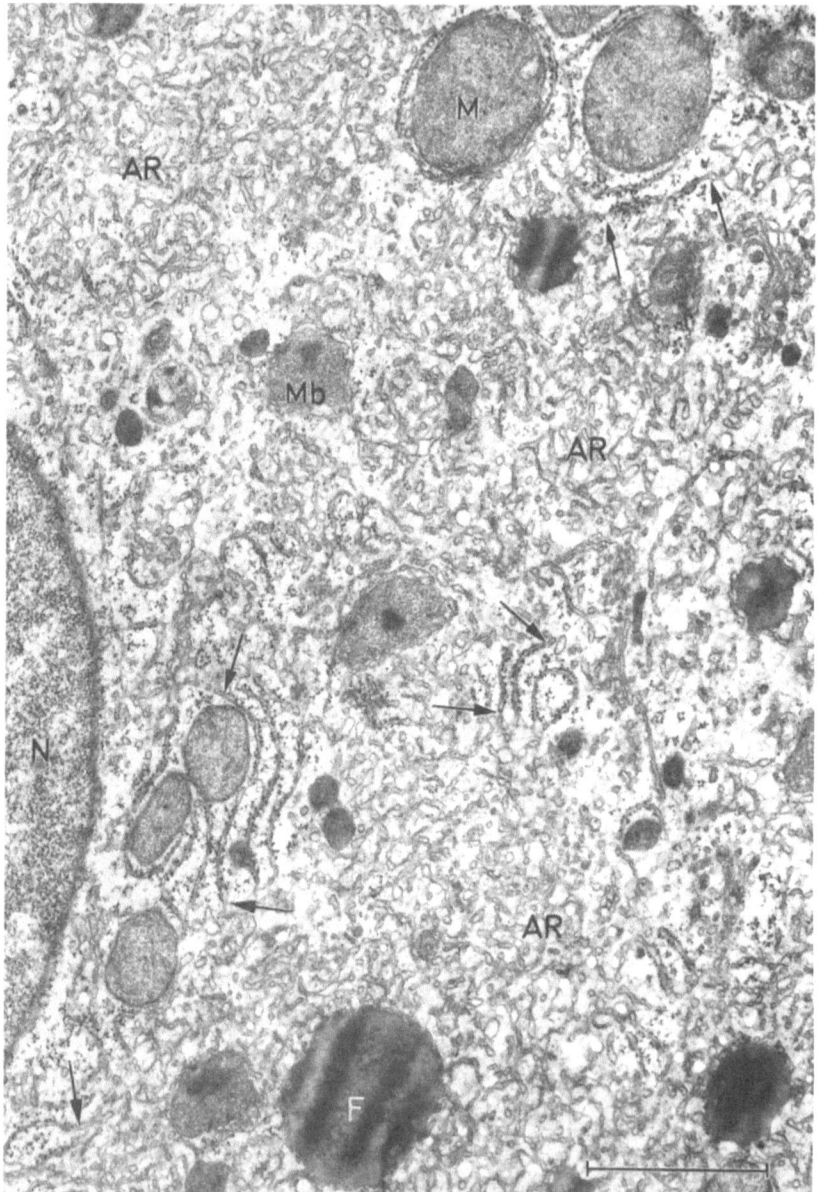

Fig. 42. Portion of a relatively glycogen-poor hepatocyte showing a "diffuse" hypertrophy of the agranular reticulum (AR). The tubular elements of the smooth network are often connected with rough cisternae ($\downarrow$). Nucleus (N). Mitochondria (M). Microbodies (Mb). Fat droplets (F). 12 mg⁰/o NNM-solution, 16 weeks. Uranyl acetate. 18 000:1

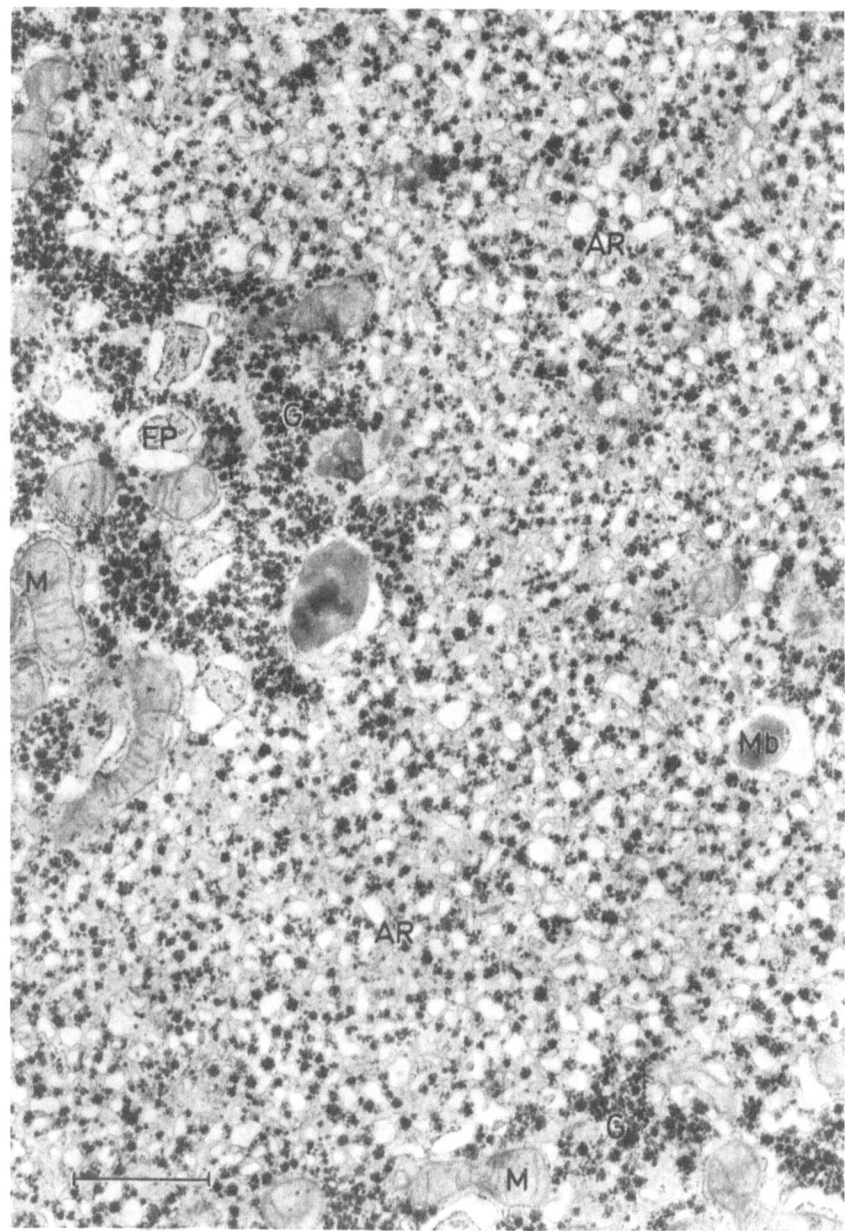

Fig. 43. Portion of a relatively glycogen-poor hepatocyte showing a "diffuse" hypertrophy of the agranular reticulum (AR). The tubular elements of the AR are somewhat dilated. In the meshes of the network dense glycogen particles, mostly forming rosettes (α particles). Some "ergastoplasm pockets" (EP) within larger glycogen accumulations (G). Mitochondria (M). Microbodies (Mb) surrounded by granular or agranular cisternae of the ER. 20 mg% NNM-solution for 7 weeks, followed by plain water for 21 weeks. Lead hydroxide. 18 000:1

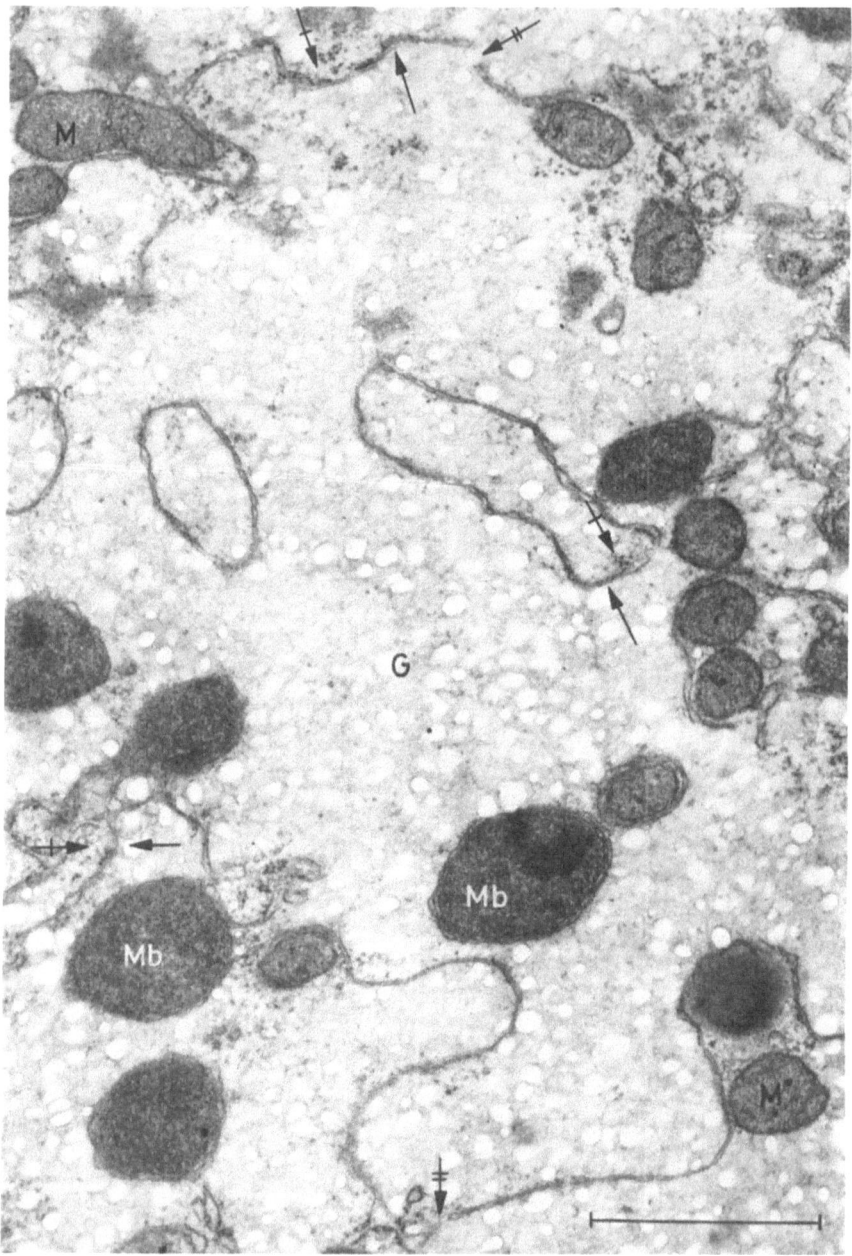

Fig. 44. Portion of a glycogen storage cell. Meandering ER-cisternae which are characterized by an unusual combination of agranular (↓) and granular (⊥) membranous elements
("combined csiternae") within a large cloudy glycogen zone (G). Sometimes the long profiles
show short interruptions (‡), obviously pores. In close contact with the ER-cisternae
mitochondria (M) and microbodies (Mb). 12 mg⁰/o NNM-solution for 12 weeks, followed
by plain water for 8 weeks. Uranyl acetate. 31 000:1

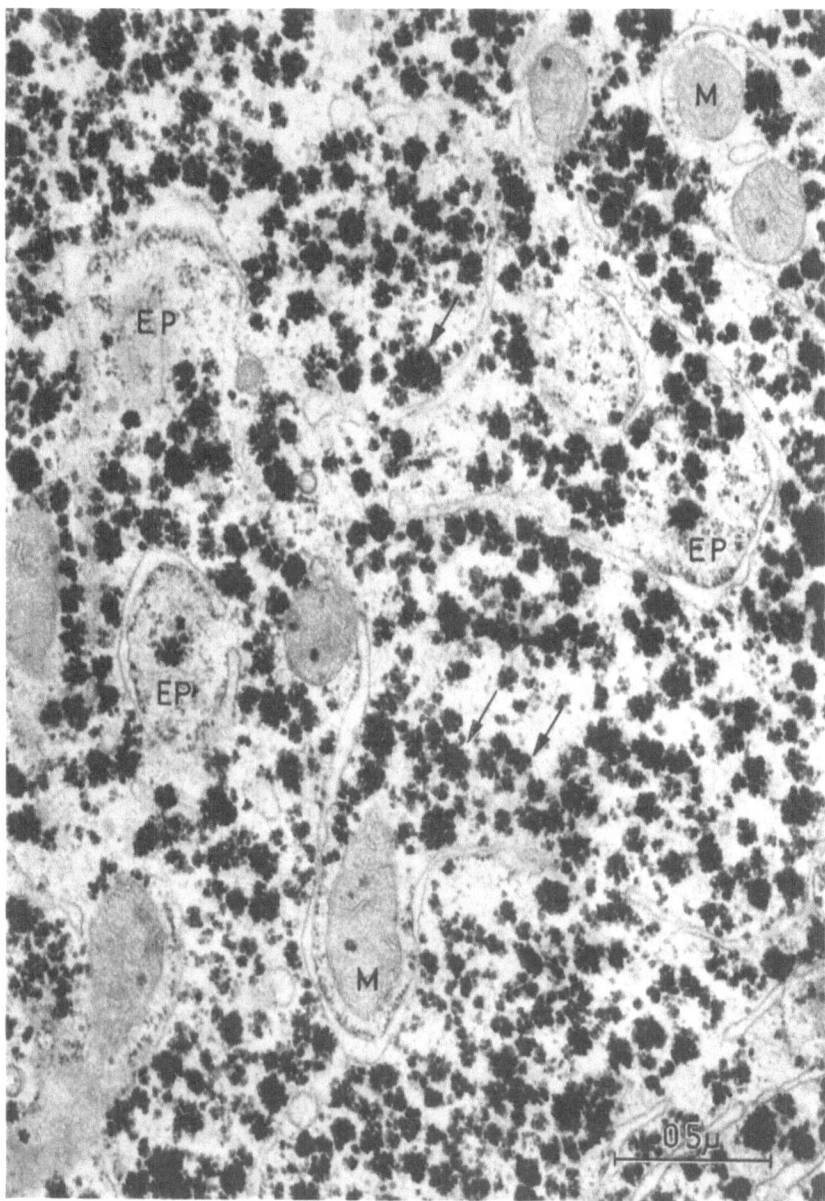

Fig. 45. Portion of a glycogen storage cell. Masses of dense glycogen particles forming mostly rosettes (↓). Long, meandering "combined cisternae" of the ER, partly smooth, partly rough, within the accumulated glycogen. The smooth parts of the cisternae are invariably in close contact with glycogen particles. The rough parts usually show inpocketing. The "ergastoplasm pockets" (EP) contain free ribosomes or mitochondria (M), while glycogen particles are rare or totally absent. 20 mg% NNM-solution for 7 weeks, followed by plain water for 21 weeks. Lead hydroxide. 42 000:1

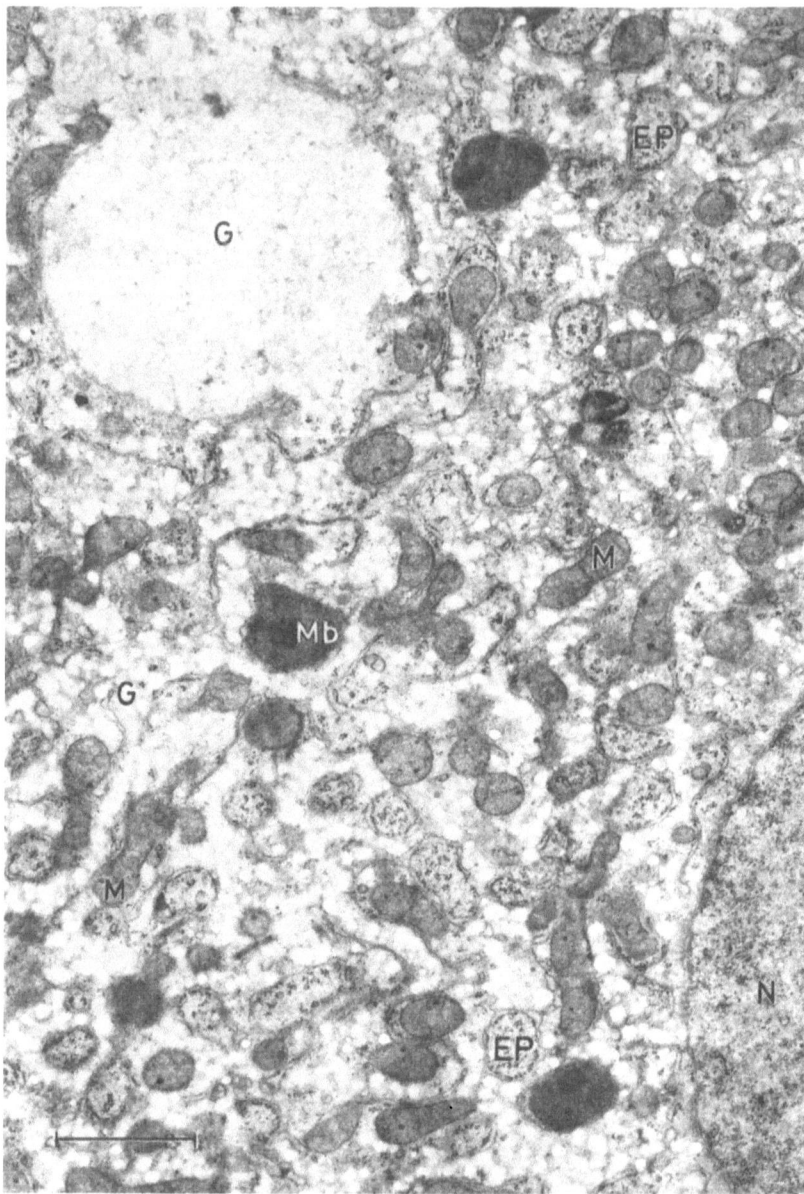

Fig. 46. Portion of a glycogen storage cell. Loosely connected "combined cisternae" of the
ER throughout the cytoplasm. Multiple "ergastoplasm pockets" (EP) enclose small glycogen-
free islands. Within the pockets one frequently observes mitochondria (M) or microbodies (Mb).
Those parts of the pockets which are in direct contact with microbodies are always smooth.
Partial segregation of a glycogen-containing (cloudy) cytoplasmic area (upper left). Glycogen
(G). Nucleus (N). 20 mg⁰/₀ NNM-solution for 7 weeks, followed by plain water for 21 weeks.
Uranyl acetate. 19 000:1

absent. The outer membrane of the pocket is on the other hand associated with glycogen particles and is free of ribosomes. The U-shaped or circular "combined" profiles are sections through the pockets.

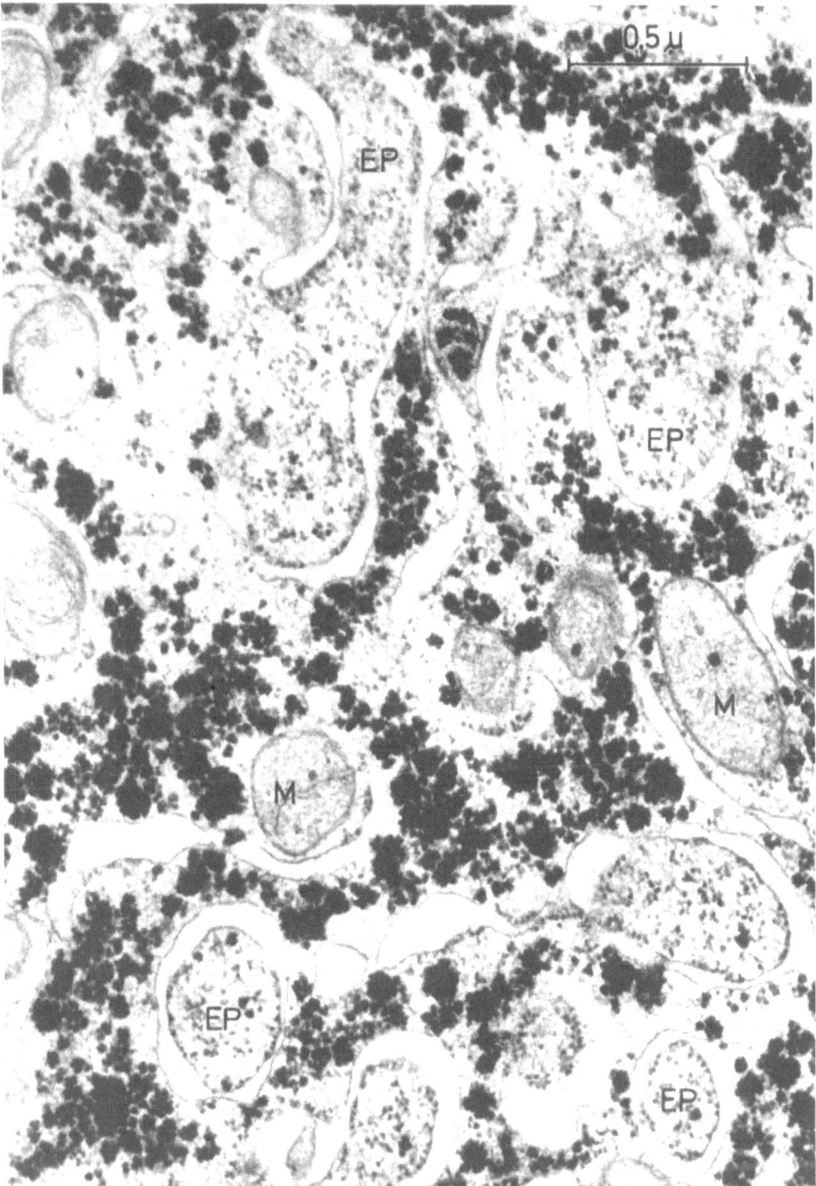

Fig. 47. Portion of a glycogen storage cell. Multiple "ergastoplasm pockets" (EP) enclose glycogen-free islands containing free ribosomes or mitochondria (M). In between the pockets many glycogen particles, mostly forming rosettes. The glycogen zones are always lined by smooth parts of the "combined cisternae". 20 mg⁰/o NNM-solution for 7 weeks, followed by plain water for 21 weeks. Lead hydroxide. 48 000:1

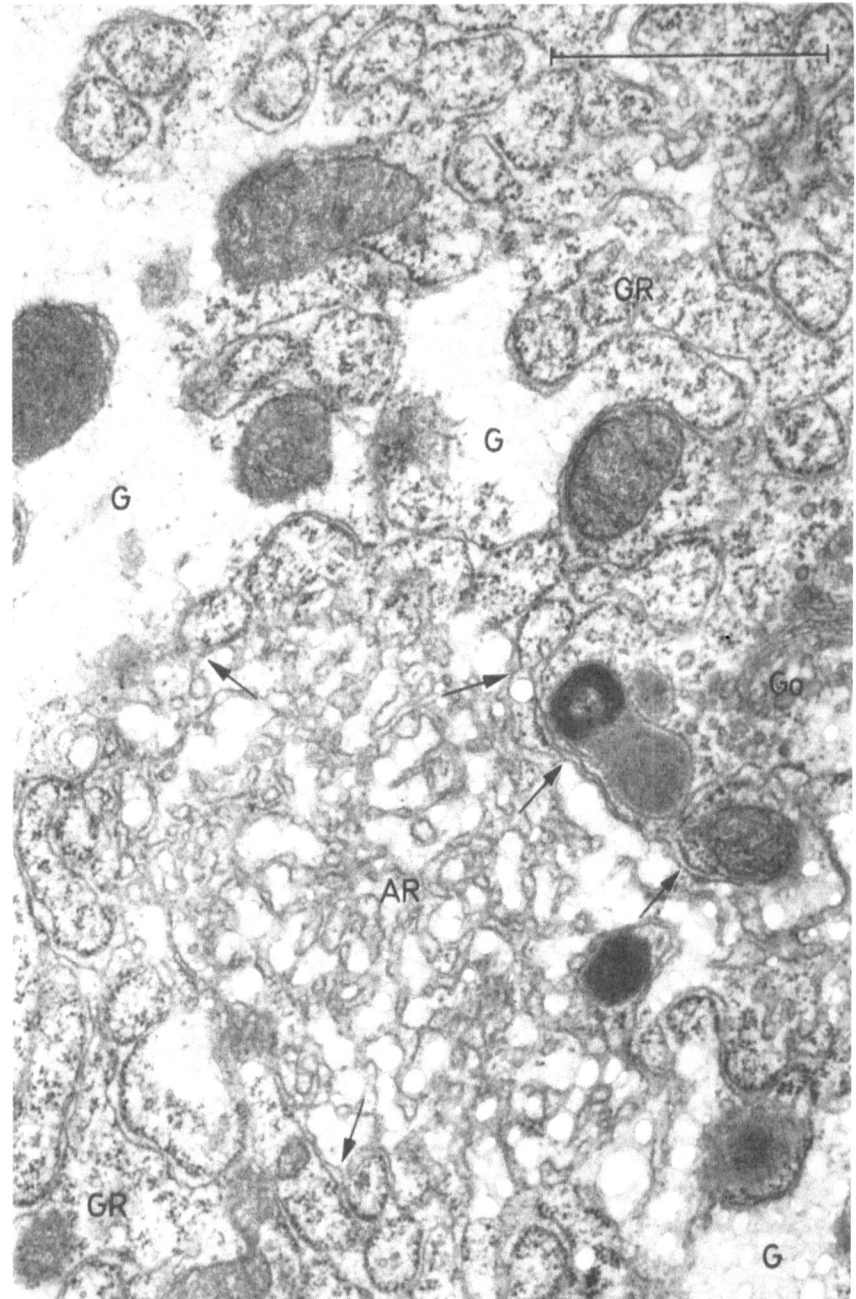

Fig. 48. Portion of a glycogen storage cell. Atypical formations of the granular reticulum (GR) showing, upon sectioning, a network of interconnected cisternae (upper right and bottom left). Between the granular arrays a typical agranular membrane complex (AR) which is at many points (↓) continuous with the granular cisternae. Glycogen (G) in the meshes of the network, at the upper left and at the bottom right. Golgi-complex (Go). 12 mg⁰/o NNM-solution for 12 weeks, followed by plain water for 8 weeks. Uranyl acetate. 37 500:1

Larger aggregates of these ER-complexes may be differently shaped: first, one observes a loose reticulum consisting of cisternae of the above mentioned type which are occasionally connected with each other (Figs. 46 and 47). Secondly, one detects a closely interwoven reticulum, partly formed by areas of the typical smooth network, partly by unusual ergastoplasm formations (Fig. 48). The second type needs a detailed description: upon sectioning, the smooth as well as the rough constituents seem to form a network in themselves or by connection with each other. One observes all stages of transition between, on the one hand, a predominantly smooth network with only a few ribosomes along the membranes and, on the other, unusual forms of ergastoplasm consisting of intercommunicating rough elements as in the case of the ergastoplasm pockets, they surround circular or oval cytoplasmic islands which contain large numbers of free ribosomes, mitochondria or microbodies. While glycogen particles may be observed within the meshes of the smooth network, there are generally none in the regions of ergastoplasm formations. Only occasionally may one encounter glycogen islands within large ergastoplasm zones. In this case, the limiting membranes of the reticulum are again smooth.

Atypical formations of the ergastoplasm similar to those just described for glycogen-storing rat liver epithelia have already been observed and analysed in detail by PARKS (1960) in the mouse salivary gland and liver parenchyma. We could, however, only rarely observe the connections with the outer membrane of the nuclear envelope described by this author. PARKS convincingly concluded from his photographs that such ergastoplasmic structures originate from multiple cylindric cytoplasmic invaginations into single rough cisternae. Their basic structure would thus be similar to that of the ergastoplasm pockets described above. The only difference would be in the greater number and length of the invaginations. Still, one cannot apply PARKS' concept directly to the interpretation of the unusual ergastoplasm formations described by us, since they are connected at many points with the typical smooth network. This peculiar phenomenon could eventually be explained, if one assumes that the smooth network complexes exhibit, at least partially, a structure similar to that of the unusual ergastoplasm formations.

One could conceive, for instance, that the typical arrangement of the smooth membranes is caused by the invagination of extremely slender — usually glycogen-containing — cytoplasmic cylinders into uniform smooth cisternae. Neighbouring cylinders would be connected by pores. This concept could explain more easily than the assumption of a three dimensional strictly tubular lattice the frequently observed transitions from smooth network formations to lamellar rough or smooth cisternae, mentioned in the previous paragraph.

In the immediate vicinity of smooth network or atypical ergastoplasmic formations one may often observe one or several *Golgi-complexes*. They consist of three to six stacked cisternae and vesicles of various size. Both cisternae and vesicles often contain a dense osmiophilic material. In most cases one cannot detect a connection between the membranes of the Golgi-complexes and the smooth or rough ER, only rarely can such connections be seen. The cisternae of the ER situated in the vicinity of the Golgi-complexes often contain osmiophilic material similar to that found inside the Golgi complexes. The cytoplasmic matrix around the Golgi-complexes is usually free of glycogen particles and ribosomes. It often seems to be more densely packed than normally.

The Chondriom

The appearance of the chondriom may vary. Some of the glycogen-rich epithelia contain relatively few mitochondria, while in others one may find a great number. They are almost invariably located in the vicinity of ergastoplasmic cisternae, a fact already described for carbon tetrachloride intoxicated rat livers by BERNHARD and ROUILLER (1956). The structure of the mitochondria frequently remains unchanged. From time to time one does observe, however, mitochondria with noticeable structural anomalies. A rarefaction of the cristae mitochondriales is the most common feature (Figs. 30 and 47), but only rarely is it so pronounced as in cells of the acinuscentral toxic pattern. The remaining cristae are often bent and may exhibit some crosslinking. In a few cases we observed a clear dilatation of the intracristal space, where, infrequently, one may detect some helix-like structures. Some mitochondria are unusually long and narrow.

Other Constituents of the Cytoplasm

Next to the above mentioned glycogen-containing cytosegrosomes, only the microbodies deserve some attention. A great number of them are found in storage cells, especially where unusual combinations of smooth and rough ER are observed (Figs. 44 and 46). We have already mentioned the localization of microbodies within the ergastoplasm pockets. We could not detect a clearcut connection between the envelope of the microbodies and the membranes of the pockets. It is noteworthy, however, that one may find a finely granulated or homogenous material within the ergastoplasm pockets which seems to be embedded in the cytoplasmic matrix without an envelope. This material is morphologically similar to the matrix of the microbodies. Further studies are needed to decide whether or not this may indicate de novo synthesis of microbodies within the ergastoplasm pockets. The microbodies are usually rather large (diameter up to 1 μ) and show a dense homogeneous matrix. Within the plane of section obtained we did not always detect a lamellar core.

II. The Transformation into Hepatomas

According to our experience the NNM-induced hepatomas always develop from parenchymal zones of enhanced glycogen storage (BANNASCH and MÜLLER, 1964; BANNASCH, 1967 a and b). Cells of the peripheral — but not those of the central cytotoxic pattern — are thus the immediate precursors of tumor cells. The beginning of pathological growth is morphologically characterized by well defined alterations in the cytoplasm of storage cells.

a) Light Microscopy

Cytological Characteristics

The transformation of a precancerous storage cell into a tumor cell is characterized by the following morphological phenomena: the *accumulated glycogen is gradually reduced* until no more glycogen can be histochemically revealed in the cytoplasm (Figs. 49—57). *At the same time one observes a net increase of the cytoplasmic*

Fig. 49

Fig. 50

Fig. 49 and 50. Glycogen-free (bright) hepatomas surrounded by glycogen-rich (dark) liver parenchyma. 12 mg% NNM-solution for 12 weeks, followed by plain water for 10 weeks. Tri-PAS. 23:1

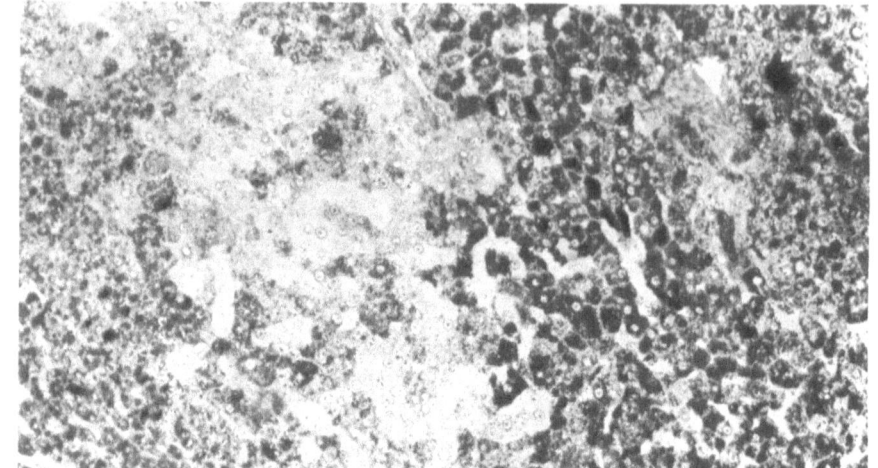

Fig. 51

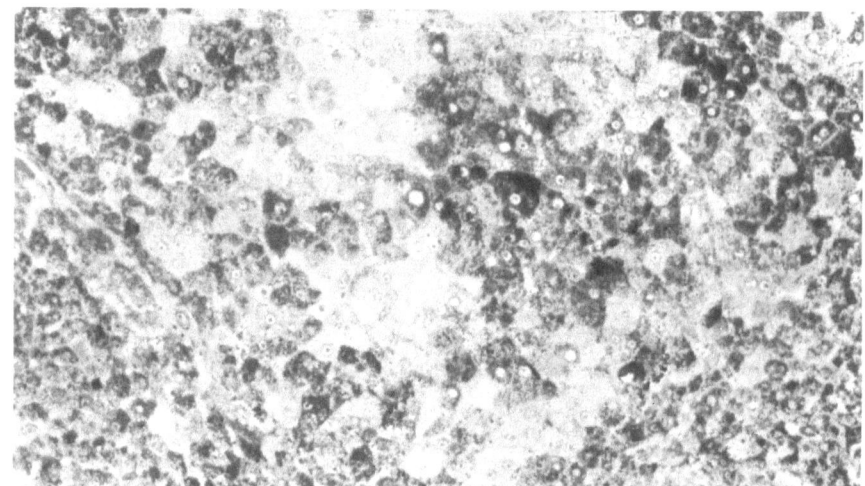

Fig. 52

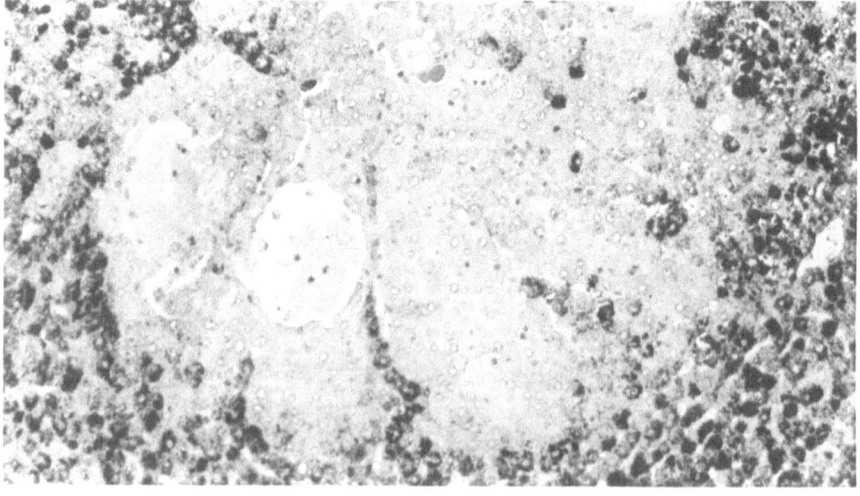

Fig. 53

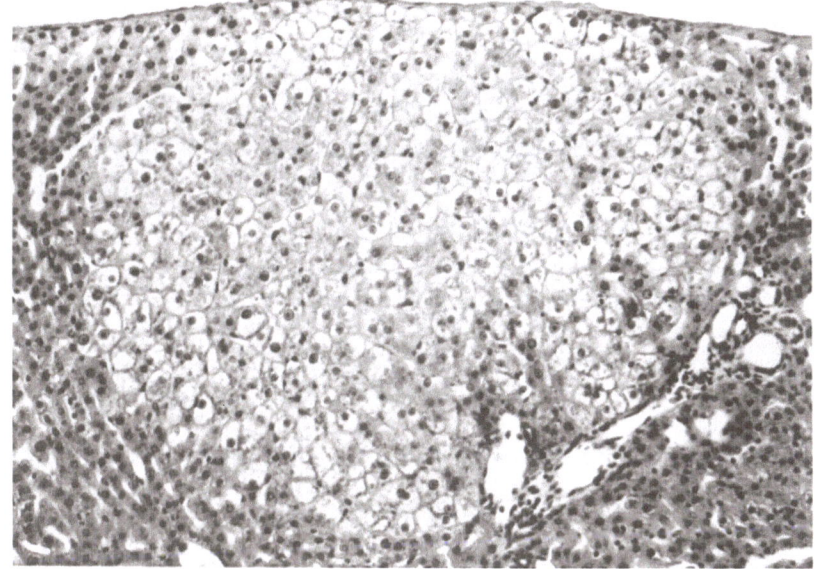

Fig. 54. Precancerous focus consisting of large clear storage cells having lost their glycogen during the preparation of the tissue. 20 mg⁰/o NNM-solution for 7 weeks, followed by plain water for 15 weeks. HE. 145:1

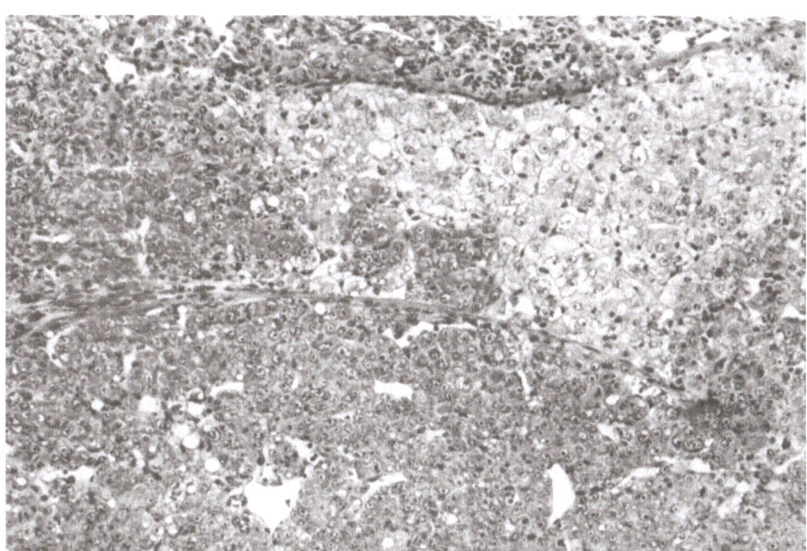

Fig. 55. Transition of a precancerous focus consisting of large clear cells (upper right) into an extensively basophilic hepatocellular carcinoma. 12 mg⁰/o NNM-solution for 12 weeks, followed by plain water for 10 weeks. HE. 145:1

Fig. 51—53. Different stages of the transformation of parenchymal foci storing excessive glycon into glycogen-free hepatomas. 51: To the right enhanced storage of glycogen, to the left beginning reduction of the stored glycogen (12 mg⁰/o NNM-solution for 12 weeks, followed by plain water for 8 weeks). 52: Advanced reduction of glycogen. In the middle and at the margin of the focus one still observes some large glycogen storage cells (12 mg⁰/o NNM-solution for 12 weeks, followed by plain water for 8 weeks). 53: Almost completely glycogen-free "microhepatoma" surrounded by glycogen-rich liver parenchyma (12 mg⁰/o NNM-solution for 12 weeks, followed by plain water for 10 weeks). All figures: Tri-PAS. 150:1

basophilia ("chromatogenesis", OPIE, 1946) (Figs. 55 and 56). Both processes may also take place in cells with a hypertrophy of the smooth ER (Fig. 57). Sometimes the metamorphosis of a glycogen storage cell into a glycogen-free hepatoma cell is accompanied by a *transitory cytoplasmic accumulation of fine or coarse fat droplets.*

Histogenesis of the Alterations as a Function of Dose and Time

The time when the first glycogen-free basophilic cells appear within the storage zones varies for experimental groups which received different doses. There may even be considerable variations among different animals within the same group. Still, one may generally say that tumor cells can be observed earlier when enhanced glycogen storage and the accompanying cytoplasmic changes also occurred earlier and were more pronounced.

After continuous administration of a *12 mg% NNM-solution* one may detect in some animals the first glycogen-free basophilic cells as early as 11 weeks after the beginning of the experiment (D∼125 mg). They can be regularly observed after 15 weeks. Metamorphosis of precancerous storage cells into cancer cells may occur simultaneously over wide areas of the organ. In many cases, however, one finds limited basophilic cell clusters, which gradually increase as a consequence of identical changes in the surrounding epithelia. The intensity of the cytoplasmic basophilia may vary considerably from one focus to another. It is noticeable that strongly basophilic areas usually consist of small cells, whereas weakly basophilic zones are formed by larger cells. One observes an increase in the mitotic rate in most basophilic areas, an increase which is generally more pronounced the stronger the cytoplasmic basophilia.

The glycogen-free basophilic cell clusters often obey the normal architecture of the liver tissue. They may, however, form unusual solid, trabecular or tubular patterns from the very beginning, thus manifesting their autonomy at an early stage. In the vicinity of such "microhepatomas" one may still find numerous liver epithelia with enhanced glycogen storage. One often observes transient zones in which the cytoplasm partly stores large amounts of glycogen or fat and partly is already devoid thereof and homogeneously basophilic. In rare cases one already observes within such areas an atypical histological pattern. It must be emphasized, however, that one never finds any signs of pathological growth, such as increased mitosis, unusual arrangement of the liver epithelia or displacement of the surrounding tissue, as long as all cells of a certain parenchymal zone show excessive glycogen storage. Glycogen-free basophilic zones, which initially do not indicate pathological uncontrolled growth either, later give rise, unmistakably, to hepatomas. This we conclude from the fact that substantial parts of the liver lobes of almost all rats which received the carcinogen for more than 19 weeks are covered by solid trabecular or tubular tumor formations. The different carcinomatous patterns are often next to one another. Usually a marginal line towards the surrounding liver tissue which still retains its normal architecture cannot be established.

Eleven out of 21 rats which continuously received a 12 mg% NNM-solution for more than 11 weeks (D∼125 mg/animal) showed multicentric hepatomas. From their histological pattern and the number of mitoses one may conclude that 7 are carcinomas (malignant hepatomas) and 5 adenomas (benign hepatomas). The diagnosis of a carcinoma has been corroborated in five cases by the detection of metastases in the lungs. One must stress, however, that the histological pattern of multicentric tumors may vary greatly from one neoplastic zone to another. For instance, we observed in

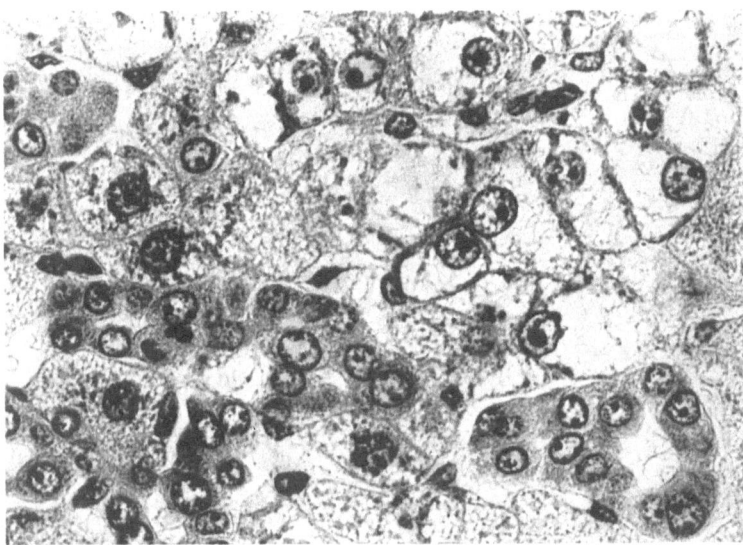

Fig. 56. Group of enlarged clear storage cells (upper right). Transition into dark hepatoma cells to the left and to the bottom (note mitoses!). 8 mg% NNM-solution, 36 weeks. HE. 700:1

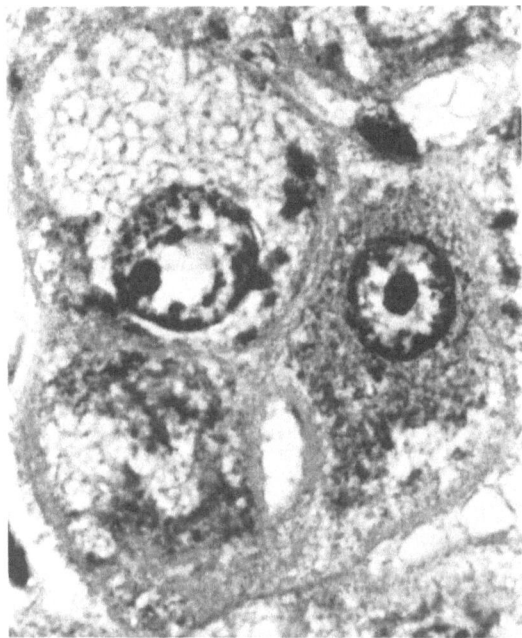

Fig. 57. Different stages of the transformation of agranular reticulum of the precancerous storage cells into ergastoplasm of the final hepatoma cells as seen under the light microscope. Upper left: storage cell containing a loose acidophilic meshwork. At the bottom left beginning and to the right advanced transformation of smooth membrane complexes into ergastoplasm. 12 mg% NNM-solution, 19 weeks. HE. 1400:1

many animals, together with definite carcinomas, tumor formations which one would rather attribute to adenomas. The histological patterns of all rat hepatomas caused by NNM correspond to those observed after administration of other carcinogens

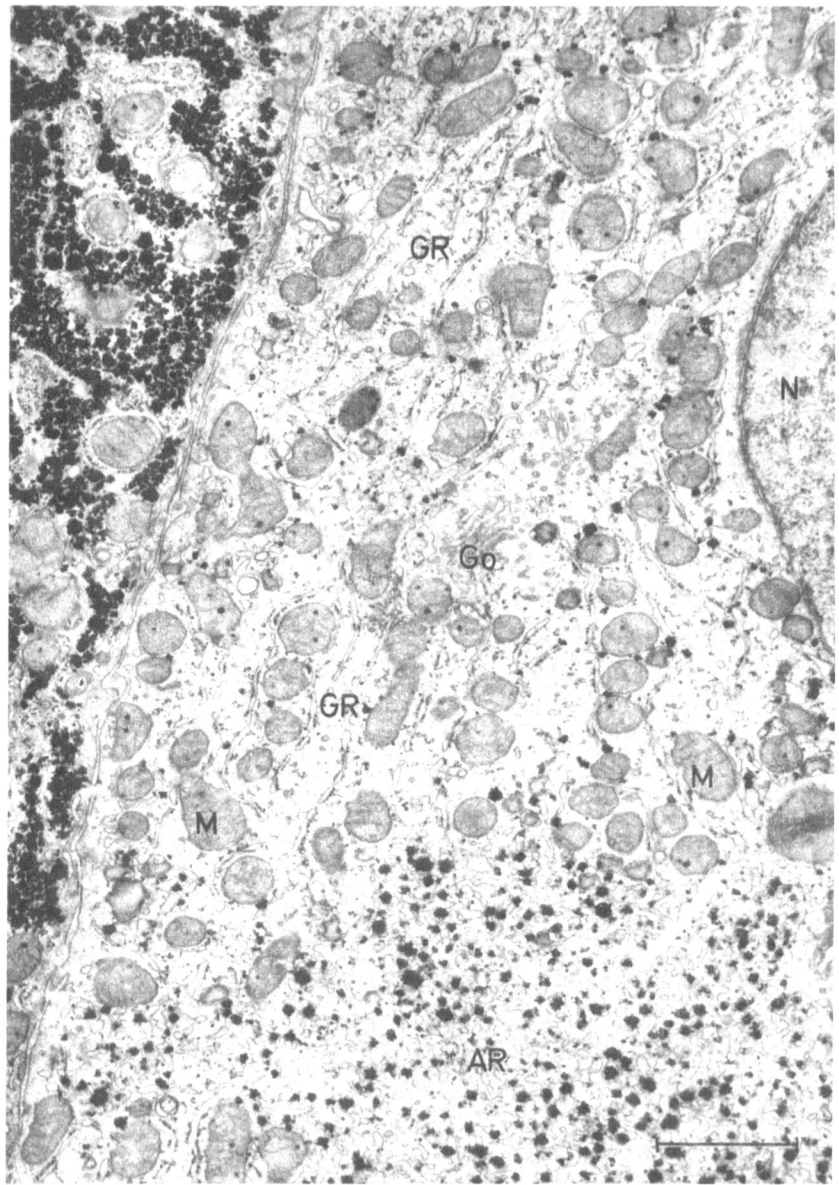

Fig. 58. Left: portion of a precancerous glycogen storage cell. Right: portion of a glycogen-poor "young" hepatoma cell containing many granular profiles (GR) and mitochondria (M). At the bottom an agranular membrane complex (AR) associated with glycogen particles. Golgi-complex (Go). Nucleus (N). 20 mg⁰/₀ NNM-solution for 7 weeks, followed by plain water for 15 weeks. Lead hydroxide. 19 000:1

(BANNASCH and MÜLLER, 1964; cf. STEWART and SNELL, 1959; THOMAS, 1961). Therefore, we need not go into any histological details.

If the carcinogenic stimulus is removed after 12 weeks *(stop-experiment)* the glycogen-free basophilic zones persist and may even increase. We cannot yet decide

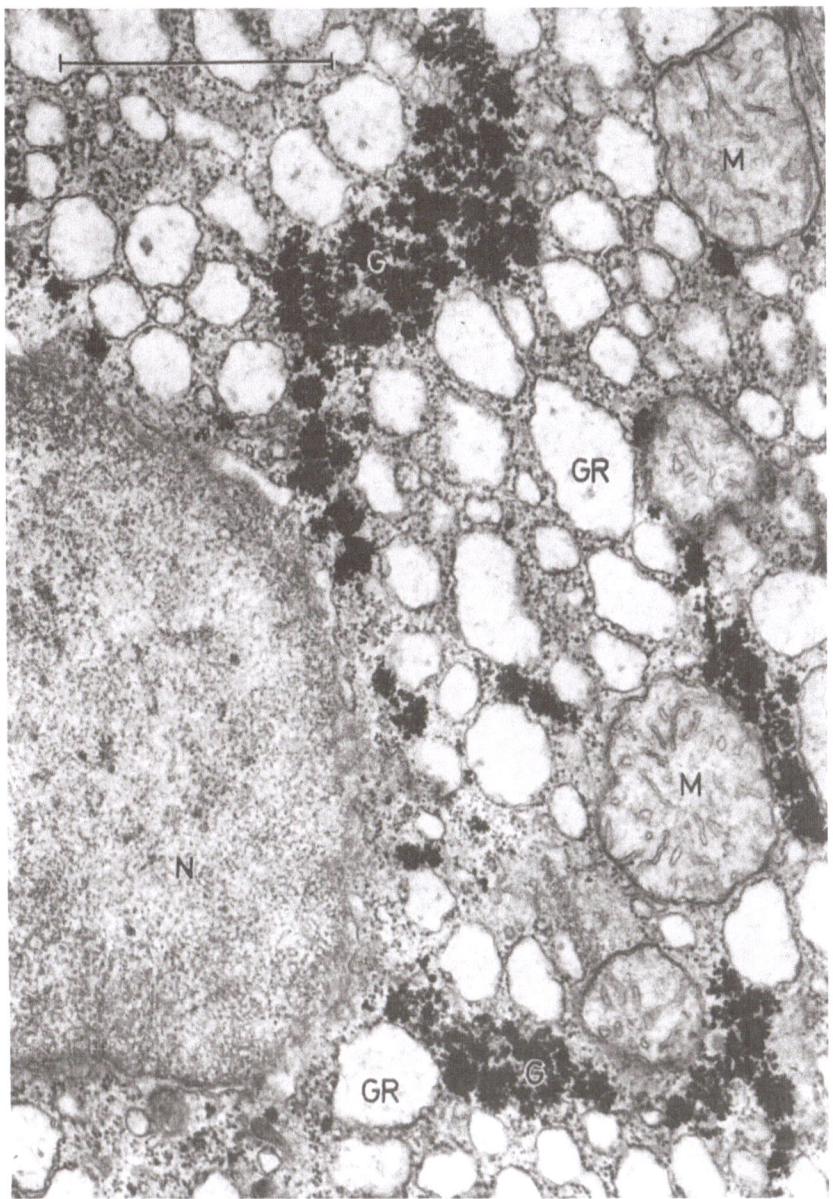

Fig. 59. "Young" hepatoma cell containing abundant vesicular ergastoplasm (GR) and free ribosomes. In addition, remaining glycogen aggregates (G). Mitochondria (M) showing some hydropic swelling. Nucleus (N). 12 mg% NNM-solution for 9 weeks, followed by plain water for 1 year. Lead hydroxide. 39 000:1

whether the enlargement of the tumor foci is only a consequence of mitotic divisions of cells that are already glycogen-free and basophilic at the time of removal of the carcinogen or are caused by a transformation of glycogen storage cells into glycogen-

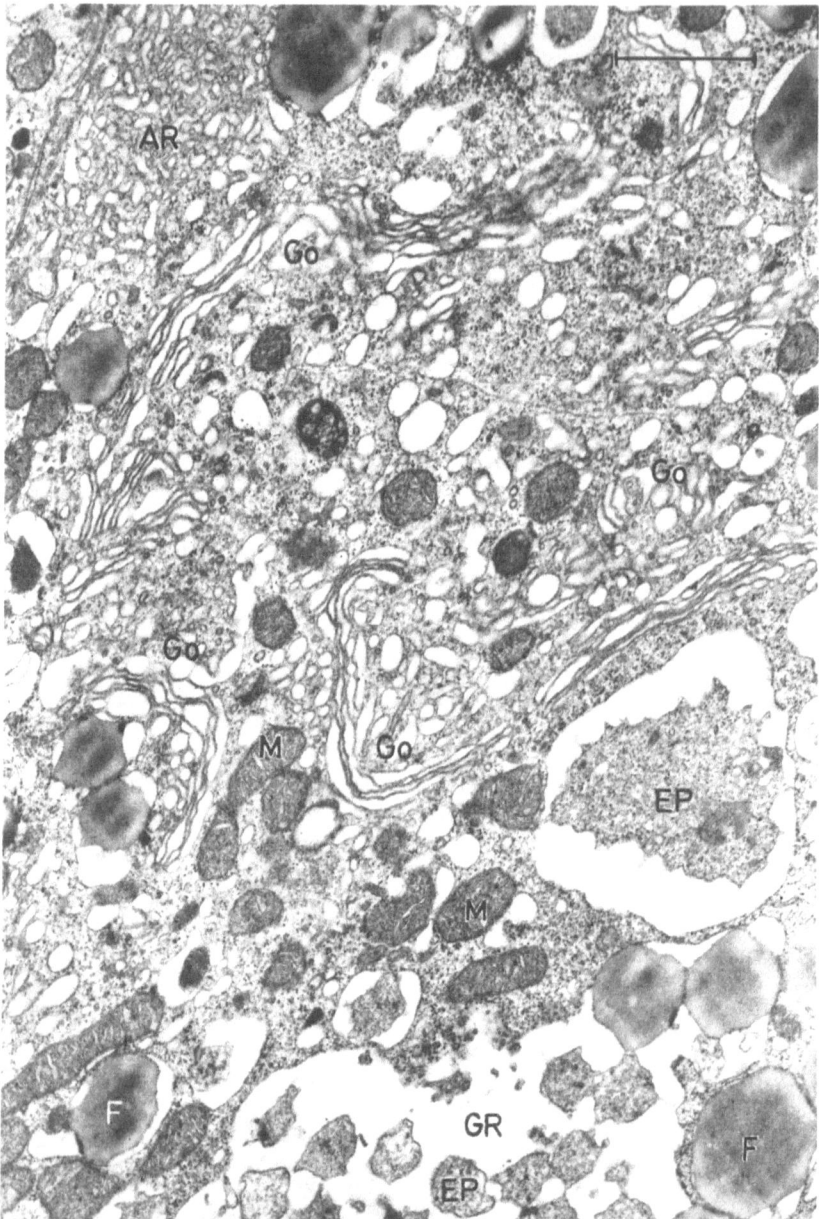

Fig. 60. Portion of a glycogen-free and ribosome-rich hepatoma cell. Bottom right: unusual ergastoplasmic formation (GR) showing multiple inpocketings (EP). Large Golgi-complex (Go) in the center. Upper left: remaining small agranular membrane complex (AR). Mitochondria (M). Fat droplets (F). 12 mg⁰/o NNM-solution, 27 weeks. Lead hydroxide. 19 000:1

free tumor cells which takes place afterwards. One certainly has to consider both possibilities, since the incidence of tumors is very high even in the stop-experiment. Sixteen out of 29 animals which were sacrificed after removal of the carcinogen (D~125—170 mg/animal) or died spontaneously showed multicentric hepatomas (8 carcinomas, 8 adenomas). Two of the animals which died spontaneously had metastases in the lungs.

After continuous administration of a *6 mg⁰/₀ NNM-solution* one observes the first glycogen-free basophilic zones 22 weeks (D~133 mg/animal) after beginning of the experiment. Thus, their development takes twice as long as in the case of a continuously administerd 12 mg⁰/₀ solution. This means that the total dose of carcinogen leading to the formation of such cells is about the same in both cases. The four rats killed later (34 and 37 weeks; D~212—246 mg/animal) all suffered from multicentric hepatocellular carcinomas, leading to pulmonary metastases in three animals.

After administration of a *20 mg⁰/₀ NNM-solution* the situation is very complex, since exposure to such high doses of toxin leads to a rapid, mostly non-specific glycogen depletion and a concomitant diffuse cytoplasmic basophilia (see page 8). It is thus very difficult to decide wheter a glycogen-free basophilic cell belongs to the centroacinal cytotoxic pattern or is a tumor cell. We assume, however, that the transformation of a storage cell into a tumor cell starts rather early, since we observed hepatomas as early as 11 weeks after the beginning of treatment with a 20 mg⁰/₀ solution (unpublished results). The significance of the various parenchymal alterations caused by the 20 mg⁰/₀ solution for carcinogenesis can only be evaluated after the removal of the toxin, that is, when the acinuscentral cytoplasmic changes have receded. The liver parenchyma now shows aspects similar to those observed after a 12 mg⁰/₀ stop-experiment: with a single exception all animals show multicentric glycogen-free basophilic foci. Seven out of 15 rats suffered from multicentric hepatocellular tumors (2 carcinomas, 5 adenomas); no metastases were observed.

One may generally say that continuous experiments lead to more malignant tumors than do stop-experiments. Details of the differences will be published later.

b) Electron Microscopy

Glycogen and Endoplasmic Reticulum (ER)

Reduction of the excessive glycogen depots as observed with the light microscope corresponds to a *progressive loss of glycogen particles* as revealed by electron microscopy. At the same time one observes an increase of free or membrane-bound ribosomes, corresponding to the increase in cytoplasmic basophilia. As already mentioned under II a (page 52), the cell may during metamorphosis pass a stage in which *fat droplets are accumulated*. Within sites of neoplastic transformation one frequently observes cells which still evidence the cytoplasmic characteristics of storage cells, such as confined aggregates of glycogen particles or large complexes of smooth ER. On the other hand, they show already fat droplets or even some of the morphological features of a typical hepatoma cell, in particular a relatively large number of ribosomes (Figs. 58 and 59). It is noteworthy, that the smooth membrane complexes are often continuous with unusual ergastoplasm formations, characteristic of certain tumor cells.

The final hepatoma cells are usually poor in or completely devoid of glycogen; only occasionally do they contain remnants of smooth membrane complexes (Fig. 60).

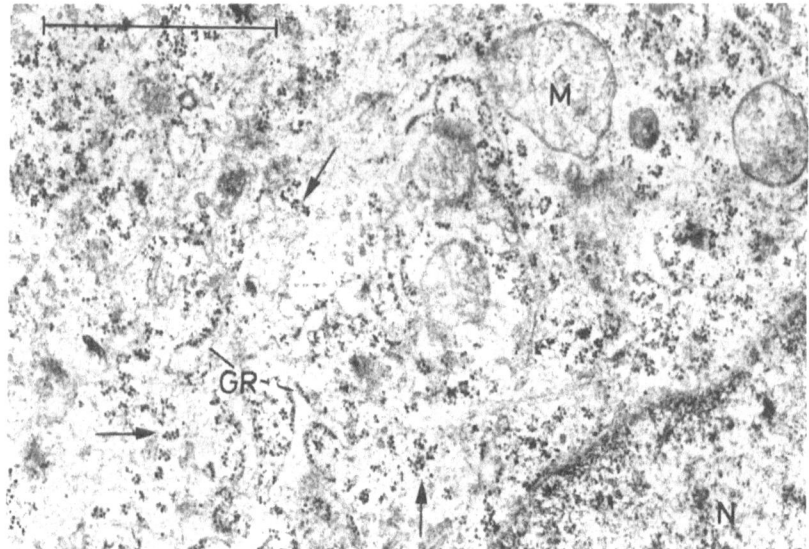

Fig. 61. Portion of a considerably de-differentiated hepatoma cell. Abundant free ribosomes forming mostly polysomes (↓). Single granular profiles (GR). Crista-poor mitochondria (M). Nucleus (N). 20 mg% NNM-solution for 7 weeks, followed by plain water for 21 weeks. Lead hydroxide. 31 500:1

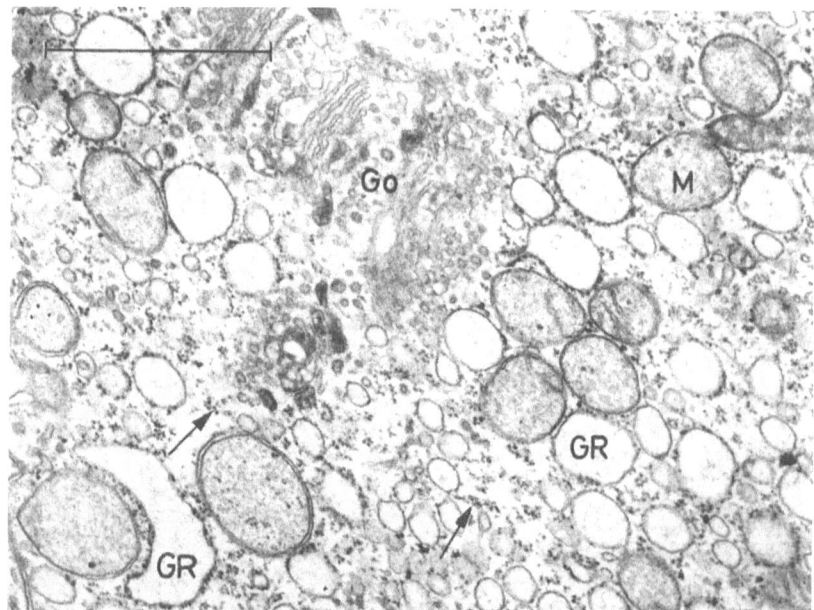

Fig. 62. Portion of a hepatoma cell. Free ribosomes (↓) together with granular vesicles (GR). Crista-poor mitochondria (M). Golgi-complex (Go). 12 mg% NNM-solution for 12 weeks, followed by plain water for 4 weeks. Lead hydroxide. 31 500:1

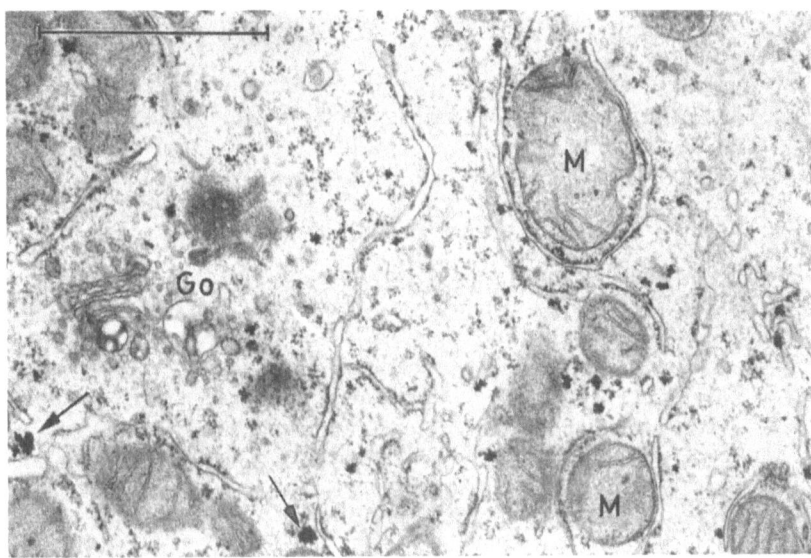

Fig. 63. Portion of a somewhat higher differentiated hepatoma cell. Together with free ribosomes one observes long, meandering ER-cisternae which have a striking similarity with the "combined cisternae" of many precancerous storage cells. Very few glycogen particles (↓). Mitochondria (M) contain relatively few cristae. Golgi-complex (Go). 12 mg⁰/o NNM-solution, 16 weeks. Lead hydroxide. 31 500:1

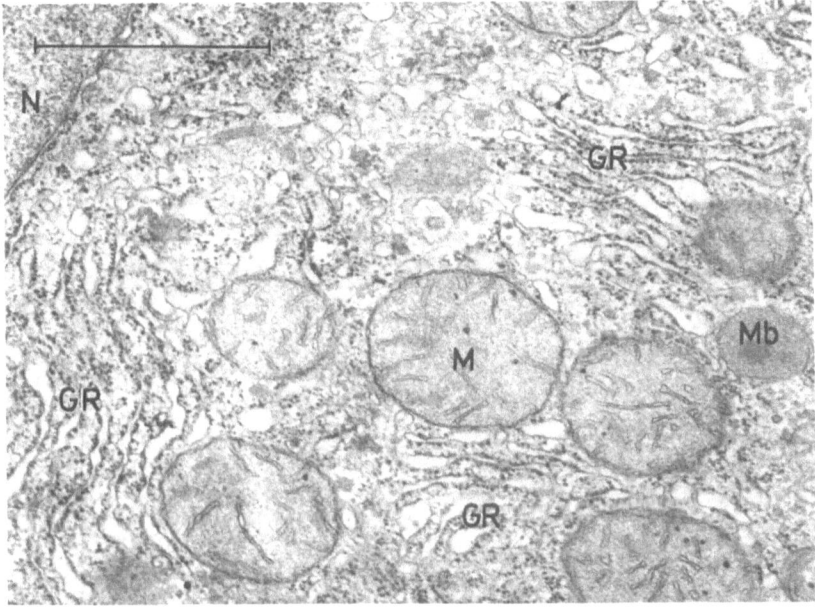

Fig. 64. Portion of a highly differentiated hepatoma cell containing abundant lamellar ergastoplasm (GR). Mitochondria (M) are morphologically intact. Microbodies (Mb). Nucleus (N). 12 mg⁰/o NNM-solution for 12 weeks, followed by plain water for 8 weeks. Lead hydroxide. 31 500:1

Besides such negative characteristics there are only a few positive ones, especially since the fine structure of the cytoplasm greatly varies from one tumor cell to another. The only ubiquitous positive feature of a hepatoma cell is the above-mentioned abundance of free or membrane-bound ribosomes. It must be stressed, however, that there are again considerable variations from cell to cell, as is the case for the cytoplasmic basophilia seen under the light microscope. Apart from merely statistical variations one observes *several distinct modes of ribosomal and ergastoplasmic organization*. This is very important, since the organization of ribosomes and ER are known to be a valuable morphological criterion for the degree of cell differentiation (PALADE, 1955; PORTER, 1961). The different patterns of these cytoplasmic components therefore indicate a different degree of de-differentiation of hepatoma cells, depending on their deviation from the normal state (cf. OBERLING and BERNHARD, 1961).

The two extremes are represented by hepatoma cells which either contain mostly free ribosomes or membrane-bound ones. The former contain only a few rough membranes (Fig. 61) which are irregularly disposed and do not form any ergastoplasm aggregates. They may be more or less dilated or very slender. When dilated they are often filled with a finely flocculated material or else seem to be optically empty. The *rarefaction of rough ER* and hence of membrane-bound ribosomes is opposed by an extraordinary *abundance of free ribosomes*. They are embedded into the matrix throughout the cytoplasm, frequently forming small aggregates (polysomes). Under the light microscope such hepatoma cells can therefore easily be discerned by their intense cytoplasmic basophilia. Such cells seem to originate mainly from epithelia with enhanced glycogen storage which contain relatively few membranes of the ER during all stages of carcinogenesis.

The intermediate type is represented by tumor cells which contain a *larger number of ribosomes together with a somewhat better developed ER* (Figs. 60, 62 and 63). The free ribosomes may be almost as densely packed as in the above mentioned tumor cell type. In many cases, however, they are just loosely scattered over the cytoplasmic matrix and one may therefore expect to see a weaker cytoplasmic basophilia under the light microscope, provided that the number of membrane-bound ribosomes is also rather small. The ER is arranged in different patterns: it partly forms the typical parallelly stacked cisternae, partly vesicles which are at some points or entirely lined with ribosomes. Finally it may be characterized by a peculiar formation which corresponds in its membrane arrangement to the unusual combinations of smooth and rough ER which can easily be recognized by their ergastoplasmic pockets and which occur in many storage cells during the later stages of carcinogenesis (Figs. 60 and 63). In the case of glycogen-free tumor cells, however, the membranes of this reticulum are usually studded with ribosomes, even outside the pockets. In addition to such ergastoplasm formations one may sometimes encounter smaller complexes of typical smooth ER (Fig. 60). The former had already been observed for other experimental hepatomas (SVOBODA, 1964; HRUBAN et al., 1965; MA and WEBBER, 1966). The above mentioned findings very clearly demonstrate the origin of such hepatoma cells from the glycogen storage cells.

Hepatoma cells with *mostly membrane-bound ribosomes* are closest to the normal liver cells, that is, they are relatively highly differentiated. Judging from the arrangement of the rough surfaced membranes there are still various degrees of differentiation within this group: the rough ER may form flat cisternae arranged in

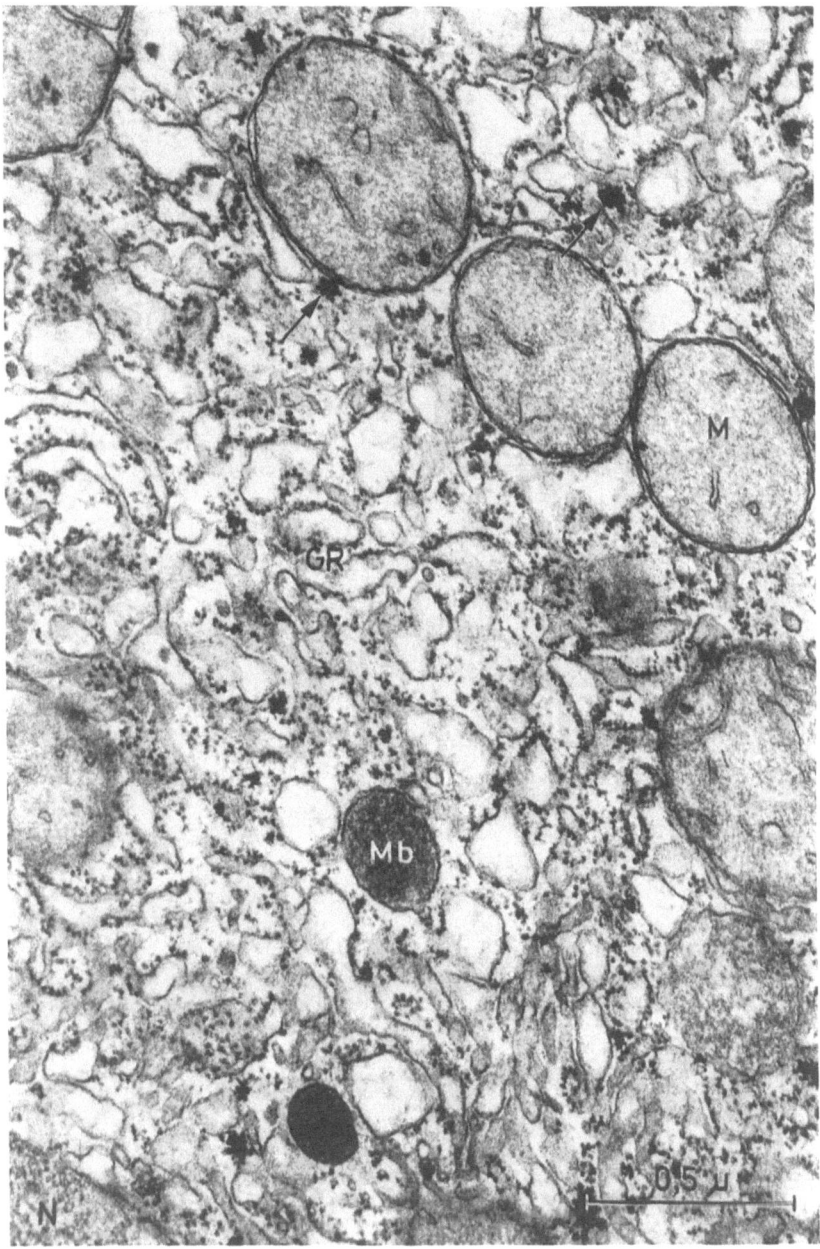

Fig. 65. Portion of a hepatoma cell. Atypical granular reticulum (GR) showing an arrangement of its membranes which is similar to that of the tubular smooth network complexes of many precancerous storage cells. Very few glycogen particles (↓) in the meshes of the network. Mitochondria (M) contain relatively few cristae. Microbodies (Mb). Nucleus (N). 12 mg⁰/o NNM-solution for 9 weeks, followed by plain water for 1 year. Lead hydroxide.
56 000:1

parallel arrays as in the case of a normal liver cell (Fig. 64) or it consists of vesicular
or tubular structures (Fig. 65). A common feature of highly differentiated hepatoma
cells is the unusual abundance of ergastoplasm. It is remarkable that the ergastoplasm
sometimes may form a tubular network similar to that observed with the smooth ER
of many precancerous glycogen storage cells (Fig. 65). Obviously, such ergastoplasmic

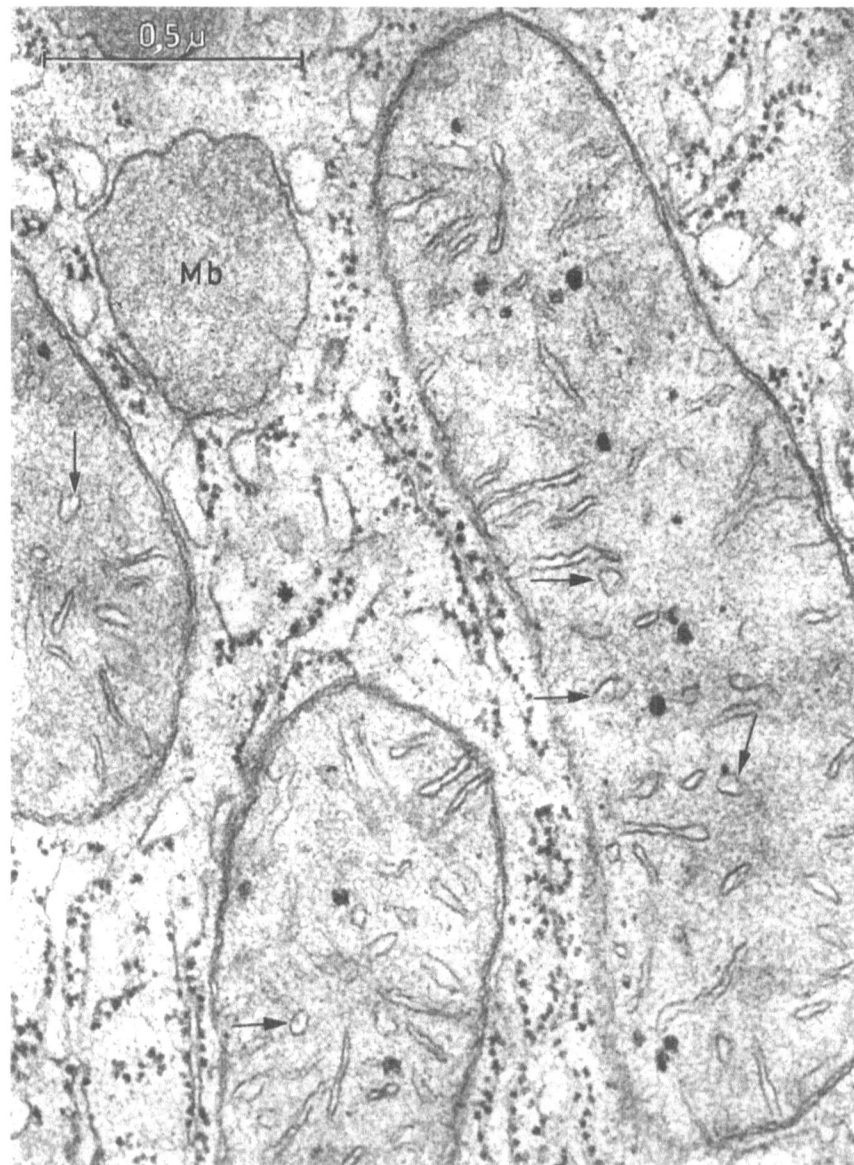

Fig. 66. Morphologically normal mitochondria af a hepatoma cell. Upon cross sectioning
the cristae often show triangular or round profiles (↓'). They occasionally contain some
osmiophilic material. Microbodies (Mb). 20 mg⁰/o NNM-solution for 7 weeks, followed by
plain water for 21 weeks. Lead hydroxide. 71 000:1

structures arise from smooth membrane complexes by the addition of ribosomes during the transformation of storage cells into tumor cells (BANNASCH, 1967 b). As yet it is not possible to decide whether this state represents the final form of the ergastoplasm of such tumor cells or whether it is an intermediate state on the way to mostly isolated flat cisternae, tubuli and vesicles.

Golgi-complexes are often found in close association with the ergastoplasm of hepatoma cells. Their basic structure is absolutely comparable to that of normal liver epithelia. They differ, however, in their greater size and larger number. Some pictures indicate that in the course of transformation of a storage cell into a cancer cell parts of the smooth ER may not only give rise to ergastoplasm, as mentioned previously, but also to Golgi-complexes (Fig. 60). Usually the final hepatoma cells contain more Golgi-complexes when they also contain more ergastoplasm membranes in the cytoplasm. Membranes of the rough or smooth ER and those of the Golgi-complexes are sometimes continuous. We did not observe the Golgi-complexes to be confined to special areas of the hepatoma cell. While they are frequently found between cell nucleus and bile canaliculus — if the latter is still present — one may also encounter them in all other regions of the cytoplasm. In one case we observed a centrosome within a Golgi-field.

The Chondriom

The appearance of the chondriom of a hepatoma cell is subject to many variations, just as in the case of precancerous liver epithelia. The mitochondria of some tumor cells differ neither in structure nor in number from those of a normal liver epithelia (Figs. 64 and 66), others show great deviations. There is not a single change in the chondriom, however, which is not matched by an opposite change in a different cell, thus losing its overall importance. We observed cells poor in mitochondria; others were rich. Small mitochondria in some cells are balanced by large ones in others. The matrix is partly loosely structured, partly rather dense. The most frequently observed structural modification is a rarefaction of the cristae mitochondriales (Fig. 61, 62 and 65). Sometimes one observes a twisting or interlacing of the remaining cristae. The same changes had been found with many liver epithelia during the precancerous phase. It is noteworthy, however, that one may sometimes encounter mitochondria in hepatoma cells which show an unusual abundance of regularly structured cristae.

Even although one may not directly correlate the pattern of the chondriom with the appearance of the ribosomes or the ER, one has to notice certain parallels. Hepatoma cells with mostly free ribosomes thus generally contain rather small mitochondria, whereas those with predominantly membrane-bound ribosomes contain larger ones. Remarkably enough, the cristae-rich mitochondria are encountered in cells with a highly differentiated ergastoplasm. Rarefaction of the cristae mitochondriales was mostly observed in cells containing a relatively large number of free ribosomes. The number of hepatomas studied seems to be too limited, however, to establish definite rules for the relationship between chondriom type and the organization of the ribosomes, that is the degree of differentiation of the tumor cell.

Other Cytoplasmic Constituents

Most tumor cells contain a more or less considerable number of dense osmiophilic bodies which correspond morphologically to the peribiliar dense bodies of a normal

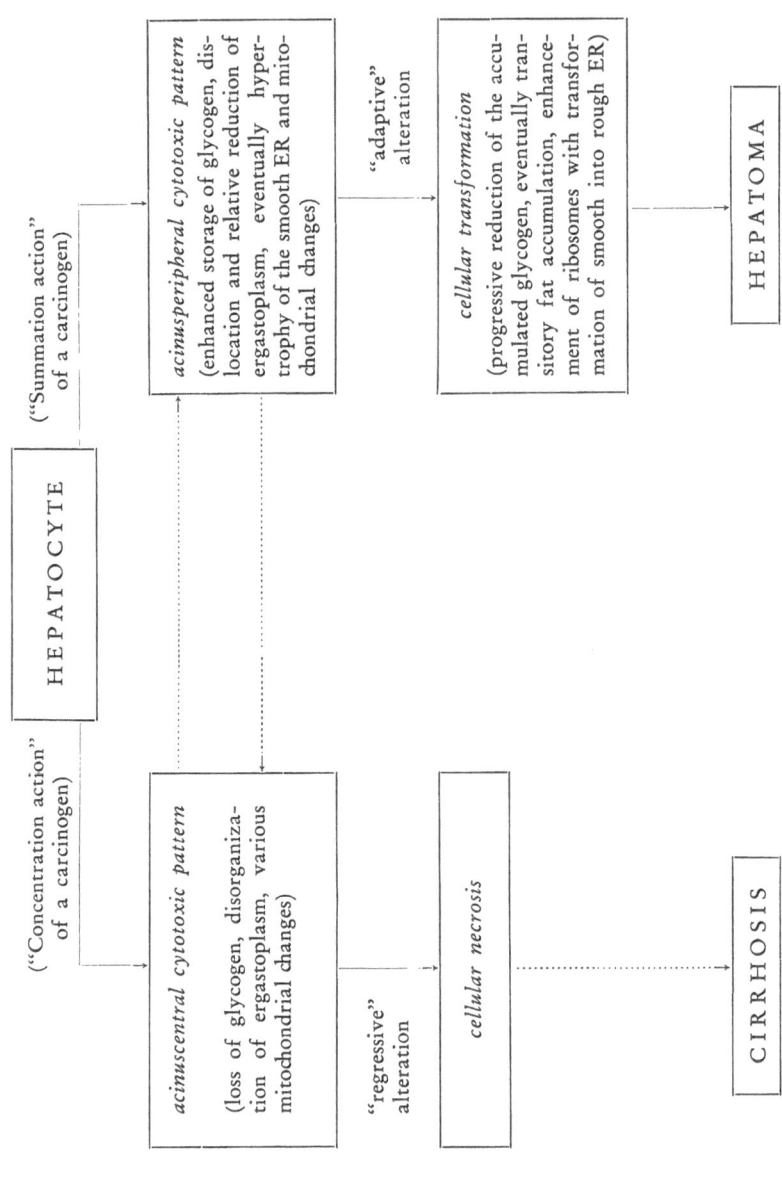

Text-figure 1. Diagram of the morphological characteristics and the significance for pathogenesis of the acinuscentral and the acinusperipheral cytotoxic pattern

liver cell. In addition, one may occasionally find cytosegrosomes of various shapes. The fact that tumor cells often contain large numbers of microbodies seems to be noteworthy. They are mainly found in cells with a high content of rough ER (Figs. 64 and 66). We could not detect any microbodies in hepatoma cells with predominantly free ribosomes. Similar observations have been reported by others (cf. de DUVE and BAUDHUIN, 1966).

D. Discussion

What conclusions concerning the role of individual cytoplasmic components can one draw from the carcinogen-induced changes in the cytoplasmic structure? Earlier studies of the NNM-intoxicated rat- and mouse-liver (BANNASCH and MÜLLER, 1964; BANNASCH, 1967 a and b) as well as the present investigations tell us that we have to start with the following premise: two basically different cytotoxic patterns can be distinguished under the influence of the carcinogen in the liver parenchyma. We called them "acinuscentral" and "acinusperipheral", according to their preferential localization within the liver lobule. Although both patterns may overlap under certain conditions to be discussed later, one should always keep them apart mentally since their "pure" forms differ not only in the structural changes of their cytoplasmic constituents but also in their fate. Their significance for pathogenesis is thus completely different. Their dose-response relationship has proved a valuable tool for discriminating between them. It is therefore necessary to discuss the relationship between dose and histotoxic pattern and the significance of the two cytotoxic patterns before evaluating the role of individual components of the cytoplasm (see text-Fig. 1).

I. Dose-dependence and Significance of the Acinuscentral and the Acinusperipheral Cytotoxic Pattern

1. Acinuscentral Cytotoxic Pattern

We have already described in detail (see "Results") how the cytoplasmic changes typical of this pattern, such as glycogen depletion and disorganization of the ergastoplasm ("chromatolysis", "diffuse basophilia") develop in only a few cells in the immediate vicinity of the central vein, when a 6 mg%/o NNM-solution is administered. If higher concentrations are used (12 or 20 mg%/o NNM-solutions) the changes affect many additional cells, starting from the center towards the periphery. Thus, restricted glycogen-free and "diffuse-basophilic" zones develop which usually occupy larger portions of the parenchyma the higher the dose (single dose). As a rule, lethal or sublethal carcinogen doses (50 or 100 mg%/o NNM-solutions) even lead to glycogen depletion and disorganization of the ergastoplasm in the entire liver parenchyma (unpublished results). It is noteworthy that the size of the affected parenchymal regions is in all cases quite independent of the total carcinogen uptake. The "stop-experiment" further proves that both glycogen depletion and ergastoplasm disorganization are usually rapidly reversible after removal of the carcinogenic stimulus. These findings support our opinion published earlier (BANNASCH and MÜL-

LER, 1964; BANNASCH, 1967 a), that *cytoplasmic changes of the central toxic pattern are due to a "concentration action" of the carcinogen,* in the sense suggested by DRUCKREY and KÜPFMÜLLER (1949). They must therefore be considered as *indications of a nonspecific cytotoxicity.* Accordingly, one cannot establish a direct correlation between such modified epithelia and tumor development by gradually following the cytogenesis and histogenesis of hepatomas.

There is, however, a close connection between the acinuscentral cytotoxic pattern and toxic liver necrosis. This is supported by different observations. First, the necrotic epithelia are generally localized within glycogen-free and "diffuse-basophilic" parenchymal zones. Experiments with different carcinogen doses further indicate that the number of necrotic cells is influenced in almost the same way by the dose of toxin as are glycogen depletion and disorganization of the ergastoplasm. Accordingly, parenchymal necroses are almost absent, if the concentration of the carcinogen is low, while they are very frequent in the case of high concentrations. The relationship between acinuscentral cytotoxic pattern and necrosis becomes even more obvious in stop-experiments: as long as large areas of the liver parenchyma undergo glycogen depletion and disorganization of the ergastoplasm under the influence of the carcinogen, one observes a high incidence of necrosis in the affected parenchymal regions. When these cytoplasmic changes recede after removal of the carcinogen necrosis becomes very rare.

Similar acinuscentral parenchymal lesions to those observed after NNM-treatment have been described for liver intoxicated with various other carcinogens. Thus it has been known for a long time that p-dimethylaminobenzene (butter yellow) leads to "chromatolysis" (OPIE, 1946), to glycogen depletion and liver cell necrosis (ORR et al., 1948 a and b) preferentially in the center of the lobule. Later, the same observations were reported for 3-methyl-4-dimethyl-aminobenzene (SPAIN, 1956; SPAIN and GRIFFIN, 1957; CHANG et al., 1958), thioacetamide (GUPTA, 1956; cf. THOENES and BANNASCH, 1962, BRODEHL, 1963), D, L-ethionine, carbon tetrachloride (SPAIN and GRIFFIN, 1957) and diethylnitrosamine (GRUNDMANN and SIEBURG, 1962). This list comprises chemically very different substances all of which lead to similar morphological changes in the central areas of the lobule, provided the carcinogen dose is kept within certain limits. Obviously, the intensity of acinuscentral parenchymal lesions is not only for NNM a function of the concentration of toxin administered, but also for other carcinogenic substances. ORR and coworkers (1948 a and b) already mentioned in their study on butter-yellow that glycogen depletion as well as liver cell necrosis are especially pronounced when the carcinogen dose is high. THOENES and BANNASCH (1962) arrived at the same conclusion using thioacetamide, a relatively weak carcinogen. They also observed that at higher doses of thioacetamide the disorganization of the ergastoplasm spreads over bigger areas of the parenchyma, thus paralleling progressive glycogen depletion. Chemical analysis of the glycogen content of rat liver after different carcinogens (SYDOW and SYDOW, 1965) and of frog liver after carbon tetrachloride intoxication (HUNTER, 1965) also led to a definite correlation between glycogen depletion and concentration of the toxin. Finally, practically the same relationships hold true for dimethyl- and diethylnitrosamine with respect to the localization and dose-dependence of parenchymal necrosis (DMNA: MAGEE and BARNES, 1956; SCHMÄHL and PREUSSMANN, 1960; DENA: SCHMÄHL et al., 1960; THOMAS, 1961; GRUNDMANN and SIEBURG, 1962).

The characteristic dose-dependence of necrosis has already led THOMAS as well as GRUNDMANN and SIEBURG to suggest that this acinuscentral lesion is due to a nonspecific "concentration action" of the carcinogen. Until now, the fact that other acinuscentral parenchymal alterations must equally be interpreted in these terms has been widely neglected. Quite on the contrary, such effects of the toxic treatment have until recently occupied a central position in discussions concerning the cytogenesis and histogenesis of hepatomas. Many authors considered glycogen depletion and "chromatolysis" (disorganization of the ergastoplasm) as important indications for a precancerous cell reaction (OPIE, 1946; ORR et al., 1948 a and b; GRAFFI and HEBEKERL, 1953; SPAIN and GRIFFIN, 1957; PORTER and BRUNI, 1959; EMMELOT and BENEDETTI, 1960; BÜCHNER, 1961; GRUNDMANN, 1961, 1962; GRUNDMANN and SIEBURG, 1962; MÖLBERT et al., 1962; OEHLERT and HARTJE, 1963; and others). There is no doubt, however, a congruence between morphology and dose dependence of acinuscentral parenchymal alterations caused by various hepatotropic carcinogens. Thus all cytoplasmic alterations of the acinuscentral cytotoxic pattern may be interpreted as a result of a nonspecific cell damage (see also CHANG et al., 1958; DAOUST and MOLNAR, 1963; HRUBAN et al., 1963; SALOMON and JEZEQUEL, 1963; STEINER and BAGLIO, 1963; LAFONTAINE and ALLARD, 1964).

Seemingly one cannot ascribe to much importance to the acinuscentral cytotoxic pattern for carcinogenesis. It does play a decisive role, however, for a completely different pathological event, namely the origin of fibrosis and cirrhosis which often develop concomitantly with hepatomas. They must be considered a consequence of toxic parenchymal necrosis. As such they are secondary effects of the acinuscentral toxic pattern. The sequence leading from a central tpye cell lesion to necrosis, fibrosis and finally cirrhosis explains why the extent of fibrotic or cirrhotic mesenchym reactions shows basically the same dose dependence as does the intensity of acinuscentral parenchymal injuries. If the latter are missing or limited to a few cells, cirrhosis or necrosis is not observed, even after several months' treatment. Such is the case after administration of a 6 mg%-solution. If the parenchymal cell damage is very pronounced, however, as after administration of a 20 mg% NNM-solution, liver cirrhosis develops very rapidly and is usually severe. All intermediate stages from light cases of fibrosis to severe cirrhosis may be observed depending on the concentration of the toxin. The acinuscentral parenchymal changes are rapidly reversible after removal of the carcinogen. It is thus easily understandable that in the case of a stop-experiment, fibrotic and cirrhotic changes generally remain stationary.

The results of our experiments just discussed support and further explain earlier opinions forwarded by THOMAS (1961), DRUCKREY and SCHMÄHL (1962) and GRUNDMANN and SIEBURG (1962). They suggested that nitrosamine-induced liver fibrosis and cirrhosis are due to the "concentration action" of such carcinogens. One may fully agree, therefore, with the conclusions drawn by the above mentioned authors, who think that these pathological processes are by no means essential for the formation of hepatomas. GRUNDMANN and SIEBURG already mentioned that the same holds true for proliferation of the bile ducts which is often observed after carcinogenous liver intoxication. *Thus, fibrosis, cirrhosis and cholangiofibrosis are no pre-requisites for the later formation of hepatomas* as had previously been suggested (cf. STEWART and SNELL, 1958). *They may merely accompany the development of such tumors.*

5*

To summarize, one may say that the *acinuscentral toxic pattern* represents a *regressive or cirrhogenic parenchymal lesion*. One has to impose two restrictions, however. First, it is still questionable whether or not all epithelia showing the acinuscentral pattern undergo necrosis; second, one should of course speak only of a "cirrhogenic" parenchymal lesion if the number of necrotic cells exceeds a certain limiting value. Individual necrotic cells as observed after administration of low doses of carcinogen are usually replaced through the mitotic activity of remaining parenchymal cells; in this case there is no increase in connective tissue (see also GRUNDMANN and SIEBURG, 1962).

We stressed several times the reversibility of glycogen depletion and ergastoplasm disorganization as well as the disappearance of necrotic cells after removal of the carcinogenic stimulus. These findings clearly support the idea that cells of the acinuscentral toxic pattern do not necessarily undergo necrosis. It would be premature, however, to conclude that the remaining acinuscentral cells are restituted ad integrum, as at least part of these epithelia turn towards the acinusperipheral pattern after removal of the carcinogen. Before going into details about the mechanism of this transformation let us first take a closer look at those cells which from the beginning show the characteristics of an acinusperipheral toxic pattern.

2. Acinusperipheral Cytotoxic Pattern

Cytoplasmic alterations typical of this pattern, such as enhanced glycogen storage, "dislocation of the ergastoplasm" and increase in cytoplasmic volume show a dose dependence quite different from that of the centroacinal parenchymal lesions. The extent of cellular lesions here is mainly determined by the total uptake of carcinogen rather than by the concentration used (single dose) (BANNASCH and MÜLLER, 1964; BANNASCH, 1967a). Consequently, they develop only after a certain latent period if low single doses are used (6 mg⁰/o NNM-solution), while they can be observed rather early in the case of high single doses (12 or 20 mg⁰/o NNM-solutions). In both cases, their intensity and the size of the affected areas increase with increasing total doses. They reach a maximum towards the end of the precancerous phase. If the carcinogen is removed from the diet after several weeks' treatment, these cytoplasmic alterations persist for weeks and months in wide areas of the parenchyma. There are, however, a few observations which imply that a short time after removal of the toxin less drastically changed cells again show a normal glycogen content and, accordingly, a normal arrangement of the basophilic bodies and a normal cell volume. This would suggest that toxic damage of the cell would have to reach a certain degree before morphological alterations of this type persist in the cytoplasm. In any case, the above-mentioned results indicate that *enhanced glycogen storage and the concomitant cytopathological phenomena in many cells are based on irreversible injuries which add up in the course of chronic intoxication.* Following DRUCKREY's and KÜPFMÜLLER's exposé (1949, see also DRUCKREY and SCHMÄHL, 1962) of the dose-response relationship of chemical carcinogens, one has to attribute the *peripheral toxic pattern to the "summation action" of NNM* (BANNASCH and MÜLLER, 1964). Microscopical evidence, however, cannot be interpreted directly as indicating a summation without loss. We cannot, therefore, decide whether or not a certain threshold has to be reached before irreversible cell damage is established. Still, we do not think, that this problem is essential to the interpretation of our morphological results. Much more important is the fact that cytoplasmic alterations of the acinusperipheral toxic pattern do show principally the same dose-dependence as can be expected for precancerous cell lesions from present-day knowledge of the pharmacodynamics of hepatocarcinogenesis (DRUCKREY and KÜPF-MÜLLER, 1949; DRUCKREY and SCHMÄHL, 1962; SCHMÄHL, 1963; RAJEWSKY et al., 1966; SCHRAMM et al., 1966). The final development of hepatomas from parenchymal foci with enhanced glycogen storage unequivocally prove that the acinusperipheral cytotoxic pattern is a precancerous lesion. During the transformation the accumulated glycogen is gradually diminished while the cytoplasmic basophilia increases concomitantly (BANNASCH and MÜLLER, 1964; BANNASCH, 1967 a and b).

There are many indications that similar conditions prevail after liver intoxication with various other carcinogens. It was already pointed out years ago that hepatomas which are caused by aromatic amines have an acinusperipheral origin (SASAKI and YOSHIDA, 1935; KINOSITA, 1937; BROCK et al., 1940; MULAY and FIRMINGER, 1952; DAOUST and MOLNAR, 1963; and others). SASAKI and YOSHIDA already described in detail how continuous administration of o-aminoazotoluene leads to the formation of "big clear cells" in intermediary and peripheral regions of the lobule. These cells form "hyperplastic nodules" from which hepatomas originate in later stages. In the course of events the big clear cells are transformed into "dark" tumor cells. We compared SASAKI's and YOSHIDA's results and pictures with ours and concluded that the big clear cells correspond to epithelia with enhanced glycogen storage, the glycogen of which was eluted during the preparation of the tissue. Our explanation is corroborated by results obtained by SPAIN and GRIFFIN (1957). They observed an enhanced acinusperipheral glycogen storage after treatment with 3-methyl-4-dimethylaminoazobenzene, a carcinogen chemically closely related to o-aminoazotoluene. They did not relate their findings to carcinogenesis, however. The "hyperplastic nodules" detected by SASAKI and YOSHIDA, which by some authors are referred to as "regenerative nodules" (see FARBER, 1963; QUINN and HIGGINSON, 1965) have since been observed after administration of almost all hepatotropic carcinogens. Most authors consider them precursors of hepatomas (OPIE, 1944; FIRMINGER, 1954; STEWART and SNELL, 1958; FARBER, 1956, 1963; QUINN and HIGGINSON, 1965; REUBER, 1965; EPSTEIN et al., 1967). The identity of NNM-induced foci of glycogen storage with the "hyperplastic nodules" is supported by FARBER's results (1963): he observed that cells within such nodules are strikingly rich in glycogen, even if the surrounding liver tissue is completely devoid of it. The same was observed by STEINER et al., (1964), EPSTEIN et al., (1967), ITO and FARBER (1967) and MERKOW et al., (1967 a and b). Enhanced glycogen storage, therefore, also seems to be responsible for such carcinogen-induced phenomena as "vacuoläre Einwässerung" or "grobblasige Zellschwellung" (BÜCHNER, 1961; GRUNDMANN, 1962; GRUNDMANN and SIEBURG, 1962; KRAMSCH et al., 1963; OEHLERT and HARTJE, 1963; KRÖGER and GREUER, 1965), "areas of hyperplasia with hydropic change" (REUBER, 1965), "hydropic degeneration" (KELLEY et al., 1966) and "hydropic swelling" (MULAY et al., 1966). This would also explain the conflict which became obvious between light and electron microscopical data. While a cytoplasmic hydropsy could never be convincingly established by electron microscopy for carcinogen-intoxicated liver parenchyma, one has often mentioned a riches of cytoplasmic glycogen. This phenomenon was observed after choline deficiency (ASHWORTH et al., 1961), thioacetamide (SALOMON, 1962; BANNASCH, unpublished results), dimethylnitrosamine (GUSTAFSON and AFZELIUS, 1963; MUKHERJEE et al., 1963), diethylnitrosamine (HAUS et al., 1964), N-2-fluorenyldiacetamide (MIKATA and LUSE, 1964), D, L-ethionine (STEINER et al., 1964), carbon tetrachloride (CONFER and STENGER, 1966) and 2-fluorenylacetamide (MERKOW et al., 1967 a and b). It is interesting to note in this context that SCARPELLI and coworkers (1963) and STANTON (1965) found the cells of carcinogen-intoxicated fish livers also to be partly very rich in glycogen. It is therefore highly probable that *enhanced glycogen storage and the concurrent cytoplasmic alterations are necessary consequences of all carcinogenic liver intoxications* (BANNASCH and MÜLLER, 1964; BANNASCH, 1967 a and b).

Similarly, the transformation of a storage cell into a glycogen free basophilic cancer cell seems to be a general phenomenon. Ever since KINOSITA's time (1937) it has often been stressed that hepatocellular tumors originate from basophilic foci (BROCK et al., 1940; OPIE, 1946; MULAY and FIRMINGER, 1952; BÜCHNER, 1961; GRUNDMANN, 1961, 1962; GRUNDMANN and SIEBURG, 1962; MÖLBERT et al., 1962; DAOUST, 1963; OEHLERT and HARTJE, 1963; DAOUST and MOLNAR, 1964). As early as 1906 LUBARSCH observed that human hepatomas lacked glycogen. EDWARDS and WHITE (1941) and PORTER and BRUNI (1959) arrived at the same conclusions studying experimental rat hepatomas. Finally, GRUNDMANN (1961) drew attention to the fact that the basophilic "microcarcinomas" are already free of glycogen. On the other hand, some transplantable hepatomas of mice and rats have been observed to be morphologically characterized by an exceptional abundance of glycogen (STRONG and SMITH, 1937; ANDERVONT and DUNN, 1952; ALBERT and ORLOWSKI, 1960; DALTON 1965; MORRIS, 1965). It is interesting that here one deals with slowly growing, low-yield transplantable tumors which belong to the "minimal deviation hepatomas". Several authors (ANDERVONT and DUNN, 1955; MIYAJI, 1965; REUBER, 1966) reported unanimously that such transplantable hepatomas lose their glycogen after several passages and at the same time grow more rapidly. The question arises therefore, whether one actually deals from the very beginning with fully developed hepatomas or whether the original transplants are only precancerous tissues or tumors in their early stages which are transformed into the actual hepatomas after several passages. In any case, close inspection of the development of NNM-induced rat hepatomas leads to the conclusion that an enlargement of the cytoplasmic glycogen depots may at best be a remnant of the precancerous phase. It is certainly not characteristic of malignancy as such.

Thus, one may characterize the peripheral cytotoxic pattern as a precancerous parenchymal lesion as opposed to the regressive or cirrhogenic central pattern (see text-fig. 1). The significance of the two different cytotoxic patterns for pathogenesis is thus clarified and we shall further be concerned with the fate of the individual cytoplasmic constituents during carcinogenesis.

II. The Fate of the Individual Cytoplasmic Components during Carcinogenesis

1. The Mitochondria

Contradictory reports exist about the fate of hepatic mitochondria under the influence of carcinogens. Some authors did not note any qualitative or quantitative mitochondrial alterations after administration of azo dyes or nitrosamines (PORTER and BRUNI, 1959; EMMELOT and BENEDETTI, 1960; BÜCHNER, 1961; HEINLEIN et al., 1962; MÖLBERT et al., 1962). Others, however, did observe various pathological alterations of the chondriom induced by the same or different carcinogens.

The most frequently encountered alteration in mitochondrial structure after exposure to carcinogens was a hydropic swelling of the "matrix type" (see THOENES, 1964) in connection with shortening, fragmentation and rarefaction of cristae. This phenomenon was observed after exposure to carbon tetrachloride (OBERLING and ROUILLER, 1956; MÖLBERT, 1957, BASSI, 1960), dimethylaminoazobenzene (TUJIMURA, 1950; HÜBNER, 1962), dimethylnitrosamine (MUKHERJEE et al., 1963), ethionine (ANTHONY et al., 1965), thioacetamide (ASHWORTH et al., 1965), N-hydroxy-2-acetylaminofluorene (HARTMANN, 1965) and isonicotinic acid hydrazide (KENDREY et al., 1967). A mere mitochondrial enlargement without hydropsy has to be distinguished from hydropic swelling. Such an enlargement was observed after choline deficiency (HARTROFT, 1962), N-2-fluorenyldiacetamide (LUSE and MIKATA, 1963; MIKATA and LUSE, 1964) and tannic acid (ARHELGER et al., 1965). In the last two cases one observed concomitant changes in shape and a rarefaction of cristae. Very striking anomalies of the inner mitochondrial membranes, such as rarefaction and deformation of cristae and dilatation of the intracristal space could be detected after exposure to thioacetamide (SALOMON, 1962 a and b; SALOMON et al., 1962) and aflatoxin B_1 (THERON, 1965). In addition to the above mentioned hydropsy the mitochondrial matrix is occasionally prone to other morphological alterations. Besides homogenous and inhomogenous densification of the matrix (TUJIMURA, 1958; ROUILLER and SIMON, 1962; THOENES and BANNASCH, 1962; BUTLER, 1966) one also detects some osmiophilic material embedded into it (WOOD, 1965). The number of intramitochondrial granulae is often reduced, in some cases they are completely absent.

The numerous older reports about quantitative alterations of the chondriom of precancerous livers or hepatomas need not be discussed in detail since they have been frequently and extensively reviewed (GRAFFI and BIELKA, 1959; LE BRETON and MOULE, 1961; NOVIKOFF, 1961; OBERLING and BERNHARD, 1961). LE BRETON and MOULE pointed out that quantitative studies are only meaningful, if they permit an estimate of the mitochondrial population of an individual cell. Analysis of different investigations with respect to this question, where possible, seems to indicate that the number of mitochondria usually decreases during carcinogenesis (PRICE et al., 1949, 1950; SCHNEIDER et al., 1953; STRIEBICH et al., 1953; ALLARD, 1957). Some electron microscopical studies — including our own — have revealed, however, an unusual abundance of mitochondria in many intoxicated parenchymal cells (ROUILLER and SIMON, 1962; SALOMON, 1962; HARTROFT, 1962; MIKATA and LUSE, 1964). Similar contradictory results have been obtained for experimental hepatomas. While most authors have observed a reduction of mitochondria in tumor cells (HOWATSON and HAM, 1955; FURUTA, 1957; HEINE et al., 1957; NOVIKOFF, 1957; TUJIMURA, 1958; DE MAN, 1960; HEINLEIN et al., 1962; SVOBODA, 1964), there are also reports of a relative increase (DJACZENKO and ALBERT, 1962; ESSNER and NOVIKOFF, 1962).

The literature quoted above clearly indicates that administration of chemically different carcinogens leads to marked changes in the fine structure and probably also in the number of mitochondria. Our own experience with NNM-intoxicated rat liver indicates that the reaction of mitochondria is not uniform throughout the liver parenchyma. The chondriom of numerous cells looks normal after weeks of exposure while others show anomalies of various kinds and extent within two weeks after the beginnung of the treatment. Similar conditions prevail after administration of thioacetamide (SALOMON, 1962) and dimethylnitrosamine (MUKHERJEE et al., 1963). One may therefore assume that such *variations in mitochondrial behaviour are a common feature of hepatocarcinogenesis*. The reasons for the inhomogenous reaction are not known. One may, however, conclude from the present results that it is not the single mitochondrion but the entire chondriom which reacts inhomogenously to chronic intoxication. While the individual mitochondrion may undergo various alterations, all mitochondria of one cell usually undergo the same changes. Thus, the chondriom of a single liver cell does react uniformly to NNM-intoxication. Similar results were reported by THOENES (1964) for the tubulus epithelia of the ischemically damaged kidney. The ability of the cellular chondriom to respond uniformly under certain pathological conditions is thus not restricted to one organ. It should be pointed out, however, that liver mitochondria may frequently react as individuals. ALTMANN (1955) arrived at this conclusion after investigations with the light microscope and his results since have been variously confirmed by electron microscopy (MÖLBERT, 1957; ROUILLER, 1960; HÜBNER and BERNHARD, 1961). It is not known why carcinogen administration tends to induce a uniform rather than an individual response in liver mitochondria.

Considering the usually synchronous reaction of the cell chondriom one is inclined to relate the heterogenous behaviour of the mitochondria during carcinogenesis to the different cytotoxic patterns observed in the carcinogen intoxicated liver. It is not possible, however, to associate distinct mitochondrial alterations with the acinuscentral or acinusperipheral toxic pattern. It is true that very pronounced alterations of the mitochondrial structure are preferentially encountered in cells of the acinuscentral pattern; basically similar anomalies may, however, be found in peripheral storage cells. It is the more difficult to systematize the results, since both "central" and "peripheral" epithelia may frequently possess a normal chondriom. Intensive investigations and a further subdivision into special cytotoxic patterns will be necessary to evaluate all details of mitochondrial behaviour during carcinogenesis.

Still, the present results permit a few important conclusions. First, *none of the qualitative or quantitative morphological alterations of mitochondria can be considered an obligatory feature of carcinogenesis,* since both the carcinogen-intoxicated liver parenchymal cell and the final hepatoma cell frequently possess a normal chondriom. On the other hand, some of the abnormal morphological phenomena observed in the chondriom of cancer cells can also be detected in many carcinogen-intoxicated parenchymal cells. One should not just discard, therefore, all morphological changes of tumor mitochondria as a consequence of secondary lesions of the final hepatoma cell, even although such secondary lesions are doubtless observed. They can, for instance, be due to a poor vascularisation of certain regions of the tumor (FASSKE and THEMANN, 1960). On the whole, one may say that the morphological states of the chondriom of precancerous livers and that of hepatomas show considerable congruence. We think, therefore, that the *various types of the chondriom are already established during the precancerous phase.* Their heterogenity would thus reflect the different history of individual tumor cells.

Such considerations do not tell us anything about the significance of the morphological alterations observed in the mitochondria of precancerous liver cells and hepatomas. It still remains to be seen whether they are causally linked to carcinogenesis or whether they merely accompany this process. This problem has been discussed for years, yet it still cannot be unequivocally answered. Many observations seem to indicate, however, that *at least the majoritiy of carcinogen-induced mitochondrial alterations is non-specific.* This holds especially true for the frequently observed hydropic swelling, a morphological phenomenon which also takes place after cell damage, which is not caused by carcinogens and which in many cases is a reversible process (cf. MILLER, 1958; ROUILLER, 1960; ROUILLER and JEZEQUEL, 1963; DAVID, 1964; STEINER et al., 1964). Similarly, enlargement of the intracristal space, density variations of the ground substance and embedding of osmiophilic material into the matrix are certainly not characteristic of a precancerous cell damage, since they are also frequently encountered in the case of non-carcinogenic lesions (cf. DAVID, 1964; STEINER et al., 1964).

The unusual mitochondrial alterations which develop mainly in acinuscentral cells after NNM-intoxication were described in detail in the experimental part. A marked decrease of cristae mitochondriales, various deformations and unusual arrangements of the remaining cristae, the occasional appearance of a transverse partition or a homogenous densification of the matrix are typical of such alterations. The diameter as well as the number of the affected mitochondria often increase. If such mitochondrial lesions were strictly limited to cells of the acinuscentral pattern, that is, cells which undergo necrosis after extended treatment with the carcinogen, one could deny any connection with carcinogenesis, just as in the case of the above-mentioned mitochondrial alterations. Occasionally, however, such lesions are also found in peripheral storage cells and in hepatoma cells (hepatoma cells: HOWATSON and HAM, 1955; HEINE et al., 1957; GRISHAM, 1960; OBERLING and BERNHARD, 1961; BRUNI, 1962; DJACZENKO and ALBERT, 1962; HEINLEIN et al., 1962; SALOMON and JEZEQUEL, 1962; TROTTER, 1963; PORTER and BRUNI, 1964; SVOBODA, 1964; HRUBAN et al., 1965, 1966; MA and WEBBER, 1966; REUBER, 1966). Similar mitochondrial anomalies were also observed in experimental rat-kidney tumors (MAO and MOLNAR, 1967) and in some human carcinomas, such as malignant cholangioma (JEZEQUEL, 1959), adenoma of the parathyroid (LANGE, 1961; HOLZMANN and LANGE, 1963), papillary cystadenoma lymphomatosum (TANDLER and SHIPKEY, 1964; McGARVAN, 1965), sarcoma alveolare molle (SHIPKEY et al., 1964), oncocytoma of the parotid (BALOGH and ROTH, 1965; HÜBNER et al., 1967), renal clear cell carcinoma (SELJELIED and ERICSSON, 1965) and hepatocellular carcinoma (THERON and MEKEL, 1964; GHADIALLY and PARRY, 1966; TOKER and TREVINO, 1966).

It is still uncertain whether or not such mitochondrial anomalies are causally linked to carcinogenesis. Against such a connection stands the fact that most of these anomalies may be produced by experimental procedures which to our present knowledge do not induce carcinogenesis. Thus one observes elongated longitudinally arranged and densely stacked cristae not only after NNM-treatment but also after ammonium intoxication (DAVID and KETTLER, 1961), extrahepatic biliary obstruction (CARRUTHERS and STEINER, 1962), experimentally induced thyrotoxicosis (GREENAWALT et al., 1962), fatty acid deficiency (WILSON and LEDUC, 1963), exposure to the non-carcinogenic azo dye 2-methyl-dimethylaminoazobenzene (LAFONTAINE and ALLARD, 1964) and copper intoxication (BARKA et al., 1966). The functional significance of such peculiar cristal alterations is not at all clear. One may assume, however, that they indicate a malfunction of the affected mitochondria. This interpretation is corroborated by results obtained with frog kidneys (KARNOVSKY, 1963) where the transformation of short transverse cristae into long and longitudinally arranged ones is paralleled by a reduction of cytochrome oxidase activity. The detection of densely stacked cristae mitochondriales in hypermetabolic skeletal muscles (uncoupling of respiration and oxidative phosphorylation) (LUFT et al., 1962) also points to a functional deficiency of such transformed mitochondria.

In addition to these peculiar alterations one observes a further cristal anomaly after extrahepatic biliary obstruction (CARRUTHERS and STEINER, 1962) and fatty acid deficiency (Wilson and Leduc, 1963), again similar to that observed after exposure to NNM. This anomaly consists of an enlargement of the mitochondria which usually concurs with a shortening and a curved arrangement of the cristae as well as a homogeneous densification of the matrix. Together with structural changes of the individual mitochondrion one observes at least a relative increase in the number of the organelles and a progressive disorganization of the ergastoplasm. In this respect Carruthers' and Steiner's results, obtained after biliary obstruction, again agree with ours. The lack of specificity of all these pathomorphological phenomena is finally demonstrated by the fact that they can be obtained with the non-carcinogenic azo dye 2-methyl-dimethylaminoazobenzene (Lafontaine and Allard, 1964). The same holds true for the appearance of a transverse partition which is often observed in mitochondria after NNM-treatment.

Fawcett (1955) was the first to observe such transverse partitions after starvation and refeeding. He thought that they indicated a beginning transverse division of the affected mitochondria. David (1964) supported this opinion (see also Wolfarth-Bottermann, 1966). Lafontaine and Allard seemed to give further weight to this hypothesis when they observed such partitions together with a marked increase in the number of mitochondria after the use of 2-Me-DAB. Even although one usually finds a striking amount of mitochondria also in NNM-treated liver cells we would rather question the above explanations, since we could not detect any correlation between the number of partitions and the amount of mitochondria in the cytoplasm. Furthermore, such intramitochondrial partitions are found predominantly in cells containing mitochondria with considerable morphological alterations such as depletion and deformation of cristae. Lafontaine and Allard emphasized the occasional separation of the two leaflets of the partition which they took as a sign for the beginning transverse division of the mitochondrion. We could never observe such a separation to lead to a complete transection of the organelle, however, but rather to the displacement of two separate chambers towards the opposite poles of the mitochondrion (Fig. 20). We tend to assume, therefore, that the partitions do not indicate a transverse division of the mitochondrion but are sign of some malfunction of the chondriom.

The reduction of cristae mitochondriales in numerous mitochondria of NNM-intoxicated liver cells is of special interest to the problem of carcinogenesis. To our knowledge, this phenomenon has never been observed so clearly in the liver parenchyma. Similar states of the chondriom have, however, been described for hepatomas (Heine et al., 1957; Bruni, 1962; Djaczenko and Albert, 1962; Heinlein et al., 1962; Trotter, 1963; Porter and Bruni, 1964; Svoboda, 1964; Hruban et al., 1965, 1966; Ma and Webber, 1966). Before discussing the functional significance of the loss of cristae one has to ask how they may disappear. Are they degraded or do they just become part of the inner membrane of the mitochondrial envelope? The latter possibility cannot be completely excluded, since one frequently observes an increase in the mitochondrial circumference; marked deformations of the remaining cristae do, however, point at a true degradation of inner mitochondrial membranes. Considering the biochemical functions of these membranes (cf. Roodyn, 1967), this would mean that those parts of the mitochondrion are degraded which mainly carry the enzymes of the respiratory chain and oxidative phosphorylation. The respiratory functions of the cell would thus be severely impaired. The increased number of mitochondria in some of the intoxicated parenchymal cells might therefore indicate a compensation of mitochondrial malfunctions.

Finally, we shall try to analyze the morphological data in the light of Warburg's well-known hypothesis which states that carcinogenesis is caused by an irreversible damage to cell respiration (WARBURG, 1926, 1962, 1966). Some of our results, such as loss of cristae, indeed seem to support WARBURG's thesis. One then would have to assume, however, that the respiration of precancerous liver epithelia or hepatoma cells with a normally structured chondriom is also impaired. Morphological evidence can certainly not disprove such an assumption, as it is not known whether or not all functional disorders of the chondriom express themselves morphologically. Metabolic studies of isolated sections of carcinogen-intoxicated livers did reveal, however, that there is no measurable decline of respiration prior to the formation of hepatomas (DRUCKREY et al., 1958; HEISE and GÖRLICH, 1964). Reduced respiration can only be observed once the hepatomas are formed. Thus it is clear that an irreversible impairment of respiratory functions of mitochondria can certainly not be the *only* reason for carcinogenesis in liver parenchymal cells. The great variations in the behaviour of an individual mitochondrion of carcinogen-intoxicated cells leave room for doubt, however, whether or not metabolic studies of liver sections can really provide an answer as to the respiration of an individual cell, much less that of an individual mitochondrion. Therefore, one may not completely exclude the possibility that one or the other cell is transformed into a cancer cell because of irreversible damage to its mitochondria. Since this seems to be rather the exception than the rule, we shall now investigate, whether or not the behaviour of other cytoplasmic constituents suggests a different mechanism of carcinogenesis.

2. Glycogen

The results of the present investigation prove once more that the transformation of a liver cell into a cancer cell occurs with characteristic alterations in the cytoplasmic glycogen content (BANNASCH and MÜLLER, 1964; BANNASCH, 1967 a and b). Various contradictory results about the glycogen content of carcinogen-intoxicated parenchymal cells as reported in the literature have already been discussed in detail. We saw that the discrepancies can easily be explained, if one takes into account the different behaviour of glycogen in cells of the acinuscentral and acinusperipheral toxic pattern. Evaluation of the significance of these different cytotoxic patterns for pathogenesis revealed that *enhanced glycogen storage of the "peripheral" epithelia indicates a precancerous cell reaction, whereas loss of glycogen in the "central" cells is a sign of a non-specific toxic cell injury.* We shall now be mainly concerned with the following two questions: 1. What causes the changes in cytoplasmic glycogen content and 2. How can one explain the contrary behaviour of glycogen in cells of the acinuscentral and the acinusperipheral toxic pattern?

Theoretically, one could conceive that enhanced glycogen storage and loss of glycogen are caused by two basically different metabolic alterations: storage could be due to enhanced glycogenesis or inhibition of glycogenolysis; the loss could be caused by the opposite processes. From experimental evidence one may conclude that enhanced glycogen storage is very probably due to an inhibition of glycogenolysis as a consequence of a toxically induced enzyme deficiency which is in most cases, or even always, irreversible (BANNASCH and MÜLLER, 1964; BANNASCH, 1967 a and b). This assumption is supported by both the persistence of enhanced glycogen storage for months after discontinuation of the carcinogen treatment and the gradual increase in cytoplasmic glycogen content after continuous administration of the toxin. The morphological congruence between experimentally induced glycogen storage cells in the rat liver and "spontaneously" appearing storage cells of human glycogenoses (glycogen storage diseases) which in many cases were proved to be due to inborn errors in metabolism (see HERS, 1964) seem to point in the same direction. *One may thus rightly assume that the carcinogen-induced enhanced glycogen storage in rat livers represents an experimentally induced glycogen storage disease.*

Arguments in support of the relationship between glycogenosis and the formation of hepatomas have already been discussed in detail. One should be wary, however, of taking every glycogen increase in the liver parenchyma as an indication of a precancerous cell reaction. Anaphylactic shock (SOOST-MEYER, 1940) or orthostatic collaps (LANGER, 1966) also give rise to increased hepatocellular glycogen depots. In contrast to the carcinogen-induced glycogenosis, however, the glycogen accumulation is here completely reversible within a short period of time. The same seems to be true for the increased hepatocellular glycogen depots observed in different species after protein deficiency (cf. SVOBODA et al., 1964, 1966; ERICSSON et al., 1966). Since for years it has been discussed whether or not protein deficiency could promote the development of hepatomas (see OPIE, 1944; MORRIS, 1954), this phenomenon is still very interesting for the problem of carcinogenesis. The question whether or not the chemically detected increase in glycogen in the rat liver after treatment with orotic acid (SICKINGER et al., 1967) is comparable to NNM-induced glycogenosis remains open, since in this case glycogen accumulation has only been observed in short-time experiments.

The design of our experiments does not allow us to decide which enzymes of glycogen metabolism are deficient. It is interesting to note, however, that GÖSSNER and FRIEDRICH-FREKSA (1964) proved histochemically the complete absence of glucose-6-phosphatase activity from focal sites of NNM-intoxicated rat livers. A lack of this enzyme gives rise to v. Gierke's glycogen storage disease in humans (see HERS, 1964). This already led us earlier to suggest that glucose-6-phosphatase deficiency might play an important role in the origin of carcinogen-induced glycogenosis in the rat liver (BANNASCH and MÜLLER, 1964; BANNASCH, 1967 a and b). Our hypothesis receives further support from the fact that reduced activity of glucose-6-phosphatase was observed also after treatment with other carcinogens (DAB or 3-Me-DAB: WEBER and CANTERO, 1955; SPAIN, 1956; ASHMORE and WEBER, 1959; FIALA and FIALA, 1959; ethionine: ANTHONY et al., 1961; EPSTEIN et al., 1967; carbon tetrachloride: BÖRNIG et al., 1962; REYNOLDS, 1963; diethylnitrosamine: GÖSSNER and FRIEDRICH-FREKSA, 1964; SCHAUER, 1966; dimethylnitrosamine: DE MAN, 1964; 2-fluorenyl-diacetamide: EPSTEIN et al., 1967; and others). Recently published results (EPSTEIN et al., 1967) are especially interesting in this context. The authors undertook comparative morphological and biochemical studies on "hyperplastic nodules" induced by ethionine and 2-fluorenyldiacetamide. These nodules regularly show a decline of glucose-6-phosphatase activity which progresses with time of intoxication, just as we observed for glycogenosis. The same results apply to a second enzyme, phosphorylase, which also plays an important role in glycogenolysis (see also NIGAM, 1965). It thus seems certain that the long-known lack of glucose-6-phosphatase (see WEBER, 1961; WEBER et al., 1963) and of phosphorylase (GORANSON et al., 1954; HADJIOLOV and DANCHEVA, 1958), observed in hepatomas, is already manifest during the precancerous phase. EPSTEIN and coworkers could further prove that the decline in the activity of these enzymes in the "hyperplastic nodules" does actually lead to an insufficiency of glycogenolysis. Thus, glycogen disappeared much more slowly from the nodules than from the surrounding liver tissue after administration of glucagon and starvation. Without such added treatment the glycogen content of the nodules is not markedly higher than that of the normal liver tissue as determined by chemical analysis. To judge from the published photographs one may explain these results by the fact that the nodules studied by EPSTEIN and his coworkers did not only contain definite glycogen storage cells but also glycogen-poor epithelia with a hypertrophy of the smooth ER and even glycogen-free hepatoma cells.

Also important for the origin of enhanced glycogen storage is the appearance of glycogen-containing lysosomes in numerous storage cells after exposure of the rat liver to NNM (BANNASCH, 1967 b). Such bodies were first detected by BAUDHUIN and coworkers (1964) for human glycogen storage disease II (Pompe's disease). According to HERS (1963), this disease is caused by a deficiency in a α-1,4-glucosidase (acid maltase) which catalyzes hydrolytic glycogen degradation. In rat livers, this enzyme is localized in the lysosomes (LEJEUNE et al., 1963). BAUDHUIN and coworkers suggested, therefore, that the above mentioned glycogen-containing bodies are lysosomes which are deficient in α-1,4-glucosidase and hence accumulate glycogen. Despite a marked congruence of the morphological phenomena, we hesitate to apply BAUDHUIN's hypothesis to the NNM-induced glycogen-containing lysosomes and to conclude that the carcinogen induces α-1,4-glucosidase deficiency. Our results rather seem to indicate that such lysosomes originate — at least in the case of NNM-intoxication — by way of a segregation of defined cytoplasmic areas, in the sense of a focal cytoplasmic degradation (HRUBAN et al., 1963) or an "autophagy" (DE DUVE, 1963). As such they would rather point to an increased activity than a deficiency of lysosomal enzymes. This possibility has already been discussed by other authors who observed glycogen-containing lysosomes ("glycogenosomes", PHILLIPS et al., 1967) in the livers of new-born mice (JEZEQUEL et al., 1963) and rats (PHILLIPS et al., 1967) and in liver cell cultures (BIEBERFELD et al., 1966). It remains to be seen whether the uptake of glycogen by cytolysosomes of precancerous storage cells is designed to remove additional glycogen or whether this process merely accompanies lysosome formation which itself is due to other reasons. In addition to the enzymes listed, that is, glucose-6-phosphatase, phosphorylase and α-1,4-glucosidase, one will have to consider other enzymes in order to elucidate the reasons for toxically induced glycogenosis (glycogen storage disease). Nothing is known, for instance, about the behaviour of amylo-1,6-glucosidase under the influence of carcinogenic substances.

In contrast to the acinusperipheral glycogen storage, the *acinuscentral glycogen depletion is usually reversible. It is probably due to an impairment of glycogen synthesis.* This tentative explanation is supported by the observation that high doses of NNM lead to an almost complete glycogen depletion throughout the liver parenchyma; after removal of the carcinogen from the diet, glycogen is reacquired, however, and may in large parenchymal areas be accumulated even beyond its original value. For the individual cell this would, by definition, mean a change from the central to the peripheral toxic pattern. This process could be explained by the assumption of a simultaneous reversible change in activity of synthesizing enzymes and an irreversible impairment of degradative enzymes of glycogen metabolism, caused by the carcinogen during the time of administration. As long as anabolism and catabolism are equally affected, the malfunction of glycogenolysis can of course not express itself in the form of enhanced glycogen storage. As soon as glycogen synthesis is restored, however, after removal of the carcinogen, glycogen is excessively accumulated in the cytoplasm of the intoxicated parenchymal cells. Glycogen cannot be degraded any more, since the deficiency of glycogenolytic enzymes persists. This manifests itself by enhanced glycogen storage. Biochemical evidence provided by SYDOW and SYDOW (1965 a and b) who studied rat livers after treatment with carcinogenic nitrosamines and azo dyes support our interpretation. They also found glycogen depletion to be reversible and in addition proved it to be paralleled by a reversible decline in glucokinase activity. This enzyme participates in the first step of glycogen synthesis in the normal rat liver by transforming glucose into glucose-6-phosphate, in the presence of ATP. It seems certain, therefore, that carcinogenic substances reversibly inhibit hepatocellular glycogen synthesis. It is possible that, in addition to glucokinase-activity, other enzymes essential for glycogenesis, such as glycogen-synthetase are also temporarily affected (see NIGAM, 1965; SIE and HABLANIAN, 1965). The reduction of glucokinase activity is of special interest, however, since a lack of glucose-6-phosphate would not only impair glycogen synthesis but the entire hepato-cellular carbohydrate metabolism, including glycolysis and the pentose-phosphate cycle. The reduction of glucokinase activity may, therefore, well be one of the main reasons for the necrosis of cells which exhibit the acinuscentral toxic pattern.

The extent of NNM-induced glycogen depletion primarily depends on the concentration of the carcinogen. This can, however, not be the only factor influencing the impairment of glycogen synthesis, since acinuscentral epithelia are mainly affected, while theoretical considerations tell us that the concentration of the toxin should be higher at the periphery of the lobule. One would rather assume that hemodynamics lead to unfavorable metabolic conditions in the central part of the lobule, or better, the third zone of the "functional" liver acinus. Here, the liver epithelia would be more easily subject to toxic lesions. Years ago, this explanation was forwarded by ALTMANN (1948) to explain the centroacinal localization of fatty infiltration of the liver parenchyma induced by low doses of hepatotoxic substances.

If the suggested explanations for the origin of glycogen depletion and enhanced glycogen storage hold true, then the dose dependence and reversibility of these cytopathological processes imply that glycogen synthesis is less sensitive to carcinogens than is glycogenolysis. We have already mentioned that glycogen depletion becomes pronounced only after exposure to relatively high doses of carcinogen and is easily reversible within 8—14 days after discontinuation of the treatment. In contrast, severe glycogenosis (glycogen storage disease) preferentially develops after administration of low doses or in stop-experiments. Continuous exposure to high carcinogen doses is thus, strangely enough, unfavorable for the development of precancerous storage cells. This may explain a peculiar result obtained by DRUCKREY and coworkers (1962). They observed that the total carcinogen dose necessary for the development of hepatomas is lower for low single doses than for high ones. They attributed this phenomenon to a "reinforcing action" of the carcinogen.

As long as the primary cellular targets of nitrosamines are uncertain, it seems premature to discuss in detail the possible ways by which NNM induces reversible and irreversible enzyme deficiencies in carbohydrate metabolism. Most authors assume that the cytotoxic or carcinogenic effects of nitrosamines are based on their reaction or that of their metabolic products with nucleic acids (cf. DRUCKREY et al., 1967). One has to consider, however, that they may equally well give rise to primary changes in nuclear as well as cytoplasmic proteins (MAGEE and HULTIN, 1962). It is thus difficult to predict whether NNM affects certain enzymes directly, as suggested by the deletion theory as originally formulated (MILLER and MILLER, 1953) whether it inhibits enzyme synthesis, or whether it impairs the complex regulatory mechanism of enzyme activity at still another point. It is true, however, that irreversible enzyme deficiencies, postulated for the carcinogen-induced glycogenosis in the liver, may at present best be explained by an inhibition of enzyme synthesis at the nucleic acid level.

Prior to a discussion of the decisive events leading to the transformation of a storage cell into a hepatoma cell we want to take a closer look at the granular and agranular ER during the precancerous phase.

3. The Granular Endoplasmic Reticulum (Ergastoplasm)

The granular or rough ER (ergastoplasm) has lately occupied a central position in the discussion of hepatocarcinogenesis (PORTER and BRUNI, 1959; EMMELOT and BENEDETTI, 1960; BÜCHNER, 1961; GRUNDMANN, 1961; COTE et al., 1962; GRUND-MANN, 1962, 1966; GRUNDMANN and SIEBURG, 1962; HEINLEIN et al., 1962; HOBIK and GRUNDMANN, 1962; MÖLBERT et al., 1962; DAOUST, 1963; DAOUST and MOLNAR, 1964; PITOT et al., 1964, 1965; BENEDETTI and EMMELOT, 1966; SIMARD und DAOUST, 1966; MÜLLER, 1967; and others). Most authors start from the premise that the ergastoplasm of liver cells is uniformly transformed during carcinogenesis in the sense of a "chromatolysis" or "disorganization". As stated above, our own studies revealed, however, that the ergastoplasm shows a distinctly different behaviour in cells of the acinuscentral in contrast to the acinusperipheral toxic pattern, in analogy to the behaviour of glycogen.

Only the regressively altered, glycogen depleted cells of the acinuscentral pattern show pronounced structural changes in their rough ER. The cisternae are no longer parallelly stacked and individual cisternae are scattered throughout the cytoplasm, often breaking and giving rise to shorter fragments and small vesicles. They frequently lose most or all of their ribosomes. In severe cases one even observes a true reduction of the ergastoplasm. It is certainly tempting to consider these alterations of the ergastoplasm as the equivalent of the light microscopically observed "chromatolysis", which OPIE first described for butter-yellow-intoxicated rat liver (BERNHARD et al., 1952; OBERLING and ROUILLER, 1956; ROUILLER, 1957; MÖLBERT, 1957; JEZEQUEL, 1958; PORTER and BRUNI, 1959; THOENES and BANNASCH, 1962; and others). Careful inspection of the descriptions and photographs offered by OPIE, BÜCHNER and GRUND-MANN and SIEBURG, among others, clearly reveals, however, that the term "chromatolysis" was not used only for the light microscopical equivalent of ergastoplasm disorganization. It did, in addition, include the phenomenon described by us as "dislocation" of the ergastoplasm, which is characteristic of precancerous storage cells. This erroneous identification of "disorganization" and "dislocation" of the ergastoplasm can easily be explained. Even though basically different, they can hardly be distinguished under the light microscope, since both lead to a reduction of the cytoplasmic basophilia in advanced stages. Still, considering that the ergastoplasm disorganization regularly concurs with glycogen depletion, dislocation, however, with

enhanced glycogen storage, one can easily differentiate between the two, even with
the light microscope (BANNASCH and MÜLLER, 1964). Electron microscopy of storage
cells further reveals that the fine structure of the rough ER in the case of ergastoplasm
dislocation remains practically unchanged, contrary to what one observes for disorganization. The ergastoplasmic bodies are merely pushed towards the periphery
of the cell or to the vicinity of the nucleus by the accumulated glycogen; their number
per unit volume cytoplasm noticeably decreases, while the cytoplasmic volume
increases. *The reduction of the cytoplasmic basophilia of precancerous storage cells
seen under the light microscope thus is not caused by the destruction of the ergastoplasm but merely by an unfavourable ratio between the cytoplasmic volume and
the number of ergastoplasmic bodies.* The term "chromatolysis" can, therefore, not be
applied in this case and one should reserve it for acinuscentral ergastoplasmic
alterations which actually are caused by a "lysis" (disorganization) of the rough ER.

Following OPIE's investigations, many authors thought that the ergastoplasm
played a decisive role in carcinogenesis, since it contains the cytoplasmic ribonucleo-
proteins and is the site of cytoplasmic protein synthesis. Studying diethylnitrosamine-
intoxicated rat livers, BÜCHNER (1961) and GRUNDMANN (GRUNDMANN, 1962, 1966;
GRUNDMANN and SIEBURG, 1962) recently suggested that the transformation of liver
parenchymal cells into cancer cells is induced by primary changes in the cytoplasmic
"structural proteins" ("cytoplasmatische Strukturproteine"). Most of the experimental
results on which they base their hypothesis are not conclusive, however, if one
considers the different behaviour of the ER in cells of the acinuscentral in contrast
to the acinusperipheral toxic pattern. The basic evidence for their hypothesis may
thus be seriously questioned. BÜCHNER and GRUNDMANN start from the assumption
that precancerous liver cells are characterized by a destruction of the ergastoplasm
(chromatolysis). The cytoplasmic RNA- and protein synthesis would thus be seriously
impaired. They consider this impairment as the reason for a gradual transformation
of the carcinogen-intoxicated cell into a cancer cell; feed-back mechanisms between
cytoplasm and cell nucleus would play an important role throughout the process of
transformation. This hypothesis seemed to be corroborated by autoradiographic
evidence, obtained by OEHLERT and HARTJE (1963) with diethylnitrosamine-
intoxicated rat livers. They reported that under the influence of the carcinogen
^{3}H-cytidine and ^{3}H-leucine incorporation is reduced, a clear demonstration of the
inhibition of RNA- and protein synthesis. OEHLERT and HARTJE (1963) pointed out,
however, that such metabolic alterations were mainly observed in acinuscentral parts
of the lobule, that is, in those regions which exhibit ergastoplasm disorganization.
This is in agreement with the results of SMUCKLER and al., (1961, 1962) who also
found a correlation between ergastoplasm disorganization and the inhibition of RNA-
and protein synthesis (see also MAGEE, 1966). They compared electron microscopic
and biochemical studies of carbon tetrachloride-intoxicated rat livers. Such examples
of comparative morphological and biochemical studies clearly demonstrate how
careful one has to be even with the interpretation of strictly biochemical results from
experiments concerned with the RNA- and protein synthesis of carcinogen intoxicated
livers. This holds equally true for the inhibition of protein synthesis or the alkylation
of proteins in nitrosamine-intoxicated rat livers (MAGEE, 1957, 1958; BROUWERS and
EMMELOT, 1960; HULTIN et al., 1960; MAGEE and HULTIN, 1962), the alkylation of
guanine in the cytoplasmic RNA-fraction after exposure to dimethylnitrosamine

(MAGEE and FABER, 1962) and the inhibition of polysome formation recently observed in NNM-intoxicated rat livers (MEYER-BERTENRATH 1967 a and b; MEYER-BERTENRATH and DEGE, 1967). All these pathological processes probably occur at least partially in the regressively altered cells of the acinuscentral toxic pattern. One could, therefore, relate them to necrosis (see also MAGEE, 1966) rather than to carcinogenesis in the liver parenchyma.

It is impossible, therefore, to draw any final conclusions as to the role of the ergastoplasm during carcinogenesis. Experiments should be designed, autoradiographical studies, for instance, to elucidate the ergastoplasmic functions in individual precancerous storage cells. Electron microscopy has already demonstrated that morphological changes in the ergastoplasmic bodies of storage cells are rare, but this, of course, does not prove functional integrity, especially since there is — as already mentioned — a marked reduction of the number of ergastoplasmic bodies per unit volume of the cytoplasm. Further studies will have to reveal whether one is dealing with an absolute or a relative reduction. If the reduction were absolute, it would still not prove that the ergastoplasm is destroyed by the action of the carcinogen. Since the ergastoplasmic fine structure of storage cells remains almost intact, we would rather tend to assume that the reduction is caused by an inhibition of de novo synthesis of the ergastoplasm.

4. The Agranular Endoplasmic Reticulum

All hepatotropic carcinogens investigated lead to a marked increase in the agranular (smooth) ER in many intoxicated hepatocytes. PORTER and BRUNI (1959) called this phenomenon a "hypertrophy" of the AR. Its significance for carcinogenesis is not clear, especially since similar alterations may be caused by non-carcinogenic substances, such as drugs (REMMER and MERKER, 1963, 1964). Our own investigations with NNM-intoxicated rat livers contribute two new aspects: first, they prove that, at least in the case of NNM-treatment, the hypertrophy of the smooth ER is always preceded by enhanced glycogen storage; secondly, they document that the additional membranes persist in numerous cells for weeks and months after removal of the carcinogen, as does toxically induced glycogenosis (glycogen storage disease). Beyond their significance for carcinogenesis these results are important insofar as they again indicate a direct or indirect functional relationship between agranular ER and glycogen in hepatocytes.

PORTER and coworkers (PORTER and BRUNI, 1959, 1960; MILLONIG and PORTER, 1960) were the first to suggest a direct functional relationship between these two cytoplasmic components which are morphologically closely associated. After studying rat livers after exposure to 3-methyl-dimethylaminoazobenzene and starvation, they concluded that the smooth ER plays an important role in glycogenesis and glycogenolysis. Their hypothesis was endorsed by numerous authors although it still lacks experimental confirmation.

Our own investigations do not provide definite evidence for the participation of the smooth ER in glycogen synthesis either. There is certainly no increase in smooth ER prior to carcinogen-induced enhanced glycogen storage. Similarly, PETERS and coworkers (1962) could not detect any augmentation of the smooth ER at the beginning or during glycogenesis in fetal mouse liver. Biochemical evidence also tends to refute a possible function of the AR in glycogenesis. Thus, one of the most important enzymes of glycogen biosynthesis, UDPG-glycogen transferase, is not associated with the smooth ER but with the glycogen particles themselves (LUCK, 1961). One cannot, of course, fully exclude any participation of the smooth ER in glycogen synthesis. Thus, ^{3}H-glucose incorporation in rat livers, studied by combined electronmicroscopy and autoradiography, recently revealed that the silver grains are found at an equal distance between the membranes and the glycogen particles (COIMBRA and LEBLOND, 1966).

Participation of the smooth ER in glycogenolysis was suggested by PORTER and BRUNI on the grounds that, parallel to the hypertrophy of the smooth ER, cytoplasmic gly-cogen depots are depleted after intoxication of the rat liver with 3-methyl-DAB or prolonged starvation. At the end of this process the former glycogen zones are occupied by densely packed membranes. The validity of this observation can no longer be questioned since similar relationships between smooth ER and glycogen have been established by numerous authors in various experiments (carcinogenic hepatotoxins: BRUNI, 1960; EMMELOT and BENEDETTI, 1960; GRISHAM, 1960; BENEDETTI and EMMELOT, 1961; EMMELOT et al., 1962; HERMAN et al., 1962; SALOMON et al., 1962; THOENES, 1962; THOENES and BANNASCH, 1962; STENGER, 1963; MUKHERJEE et al., 1963; ARHELGER et al., 1965; KENDREY et al., 1967; non-carcinogenic hepato-toxins: HRUBAN et al., 1963, 1965; STEINER and BAGLIO, 1963; LAFONTAINE and ALLARD, 1964; drugs: REMMER and MERKER, 1963; ERICSSON and ORRENIUS, 1966; JONES and FAWCETT, 1966; CHIESARA et al., 1967; LANE and LIEBER, 1967; starvation: MILLONIG and PORTER, 1960; PORTER and BRUNI, 1960; DE MAN and BLOCK, 1966). In the case of chronic NNM-intoxication the hypertrophy of the smooth ER is again frequently accompanied by a gradual reduction of initially enlarged glycogen depots. This process may obviously come to a halt at various points; in addition to the two possible extremes, glycogen storage cells with a small number of membranes and glycogen-poor cells with a great abundance of smooth ER, one frequently observes epithelia containing large complexes of smooth ER together with considerable glycogen depots. A similar combination of enhanced glycogen storage with a hyper-trophy of the smooth ER was observed by STEINER and coworkers (1964) after ethionine treatment. We cannot agree with these authors, however, when they say that their results refute PORTER's and BRUNI's hypothesis, which suggested an inverse correlation beween smooth ER and glycogen. In all cases one must consider the pos-sibility that the original level of cytoplasmic glycogen is much higher before the beginning of new membrane synthesis. *Our own results even suggest that an enhauced glycogen storage is a necessary prerequisite for a considerable hypertrophy of the smooth ER* (BANNASCH and MÜLLER, 1964). This would also be consistent with the fact that the hypertrophy of the AR is especially pronounced after administration of hepatotropic carcinogens, all of which we believe to induce focal hepatic glycogenosis. On the other hand, starvation (FAWCETT, 1955; MILLONIG and PORTER, 1960; PORTER and BRUNI, 1960; HERDSON et al., 1964; DE MAN and BLOCK, 1966) or partial hepatectomy (JORDAN, 1964) only give rise to a faint hypertrophy, in agreement with the fact that here the initial glycogen level is probably normal. As yet we do not know whether the considerable hypertrophy of the smooth ER observed after exposure to non-carcinogenic hepatotoxins or drugs is also preceded by an unusual abundance of glycogen. It is true that STEINER and BAGLIO (1963) as well as LAFONTAINE and ALLARD (1964) did emphasize the high glycogen content in the cytoplasm at the beginning of smooth membrane formation after treatment with the non-carcinogenic toxins α-naphtyl-isothiocyanate and 2-methyl-dimethylaminoazobenzene, respectively. They did not specify, however, whether the glycogen content is abnormally high or still within the normal limits. It is certainly difficult to solve this question experimentally, since both toxins give rise to a rapid glycogen depletion throughout the liver parenchyma, together with membrane formation. Any enhanced glycogen storage could thus only be *temporary* and its detection would only be possible during the early stages of intoxication.

The fact that an increase of the smooth ER usually concurs with a reduction of the cytoplasmic glycogen content does not mean that the reverse is also true. Studies with livers of new-born mice (JEZEQUEL et al., 1965) and rats (PHILLIPS et al., 1967) have unanimously shown that hepatocellular glycogenolysis within the first 12 hours after birth does not lead to an increase in agranular reticulum. The hepatic cell thus seems to be capable of degrading large amounts of glycogen without additional membrane synthesis. This does not rule out any functional relationship between smooth ER and glycogenolysis, but it provides reason for doubt.

Our own results did not remove such doubts about the validity of the above-mentioned hypothesis either. Morphological evidence of a correlation between hypertrophy of the smooth ER and glycogen depletion would only prove PORTER's and BRUNI's ideas, if it were certain that glycogen depletion is a consequence of extensive glycogenolysis. It is irrelevant in this case whether the resulting glucose is taken up by the blood stream or processed intracellularly. As long as there is no such certainty, however, one also has to consider the possibility that the glycogen content in the cytoplasm decreases because of an inhibition of glycogen synthesis. Our results would be consistent with this assumption, since it could easily explain why a cell with enhanced glycogen storage, obviously due to impaired glycogenolysis, could become glycogen-poor in the course of additional membrane formation. The inverse relationship between smooth ER and glycogen would thus indicate an indirect functional relationship between the two cytoplasmic components rather than a direct one. For instance, the additional membranes might serve to compensate deficiencies in carbohydrate metabolism.

In addition to suggested relationship between AR and glycogen metabolism, there is no doubt that the smooth membranes play an important role in other cellular processes (see PORTER, 1961; FAWCETT, 1965) Thus, the smooth ER of hepatocytes has been associated with the storage of unwanted metabolic products (STEINER and BAGLIO, 1963) cholesterol synthesis (JONES et al., 1965) and bile production (FAWCETT, 1965), functions which have not been unequivocally proved. It is certain, however, that the hepatocellular AR plays an important role in the detoxification of drugs. REMMER and MERKER (1963, 1964) were the first to demonstrate that the induction of drug-metabolizing enzymes, such as oxydases and hydrolases, after exposure of rats to phenobarbital is accompanied by a marked hypertrophy of the smooth ER. The same holds true for other drugs (HERDSON et al., 1964 a and b; ERICSSON and ORRENIUS, 1966; ORRENIUS and ERICSSON, 1966; JONES and FAWCETT, 1966; CHIESARA et al., 1967). It is therefore highly probable that the increase in smooth ER after administration of carcinogens is also associated with an increased activity of oxydases and hydrolases. It is thus tempting to explain the carcinogen-caused increase in smooth ER in terms of an induction of drug-metabolizing enzymes. This could hardly explain, however, why the increase in smooth ER persists in many cells for months after the removal of the carcinogen or why other toxically injured parenchymal cells do not show any increase. This makes us think that the *induction of drug-metabolizing enzymes accompanies rather than provokes the hypertrophy of the smooth ER in carcinogen-damaged liver cells*. It is interesting to note that the relationship between smooth ER and glycogen metabolism had been previously connected with the drug-induced hypertrophy of the smooth ER. FOUTS (1962) emphasized this point by saying that there is no known drug that induces enhanced enzyme activity without a concomitant glycogen depletion in the parenchymal cells. He even argued as follows: "If glycogen is removed in such a way as to cause changes in smooth surfaced endoplasmic reticulum there will be an effect on the microsomal enzymes". This indicates that FOUTS also considers enzyme induction by drugs to be a consequence of the hypertrophy of the smooth ER, which in itself is caused by changes in hepatocellular glycogen metabolism.

In spite of numerous parallels in the behaviour of the smooth ER caused by carcinogens and non-carcinogenic hepatotoxins or drugs, one does not know whether or not the alterations are identical. One important difference seems to be that the hypertrophy of the smooth ER persists for many months after removal of the carcinogen as we have shown, whereas it is rapidly reversible after discontinuation of drugs (HERDSON et al., 1964; ORRENUIS and ERICSSON, 1966). REMMER (1965) pointed out that one of the characteristics of drug-induced hypertrophy of the smooth ER is the lack of obvious concomitant alterations of other cytoplasmic organelles. Our own results with NNM lead us to question this statement somewhat, since we also observed a "selective" hypertrophy of the smooth ER, if for once we do not consider the initial enhanced glycogen storage. We therefore assume that the morphological alterations of the ergastoplasm or the mitochondria which frequently accompany the hypertrophy of the AR after intoxications are established only after the increase of the smooth membranes. In the case of ergastoplasm disorganization this necessarily follows from studies on ER-biogenesis which clearly established that the smooth ER is synthesized by the ergastoplasm (DALLNER et al., 1964 a and b). Hypertrophy of the smooth ER thus depends on an intact ergastoplasm. This could hardly be the case if the latter were extensively disorganized.

To summarize, one may say that the hypertrophy of the agranular ER is neither a specific nor an obligatory feature of the precancerous liver parenchyma. It is merely facultative and may to our present knowledge best be explained as the consequence of an impaired carbohydrate metabolism.

III. The Transformation of Glycogen Storage Cells into Hepatoma Cells

The transformation of glycogen storage cells into hepatoma cells always concurs with characteristic alterations of the cytoplasm (BANNASCH and MÜLLER, 1964; BANNASCH, 1967 a and b). Two cytopathological processes, glycogen reduction and increase in the number of ribosomes play a crucial role. Their development is intimately related; there may by an intermediary phase of fat accumulation. The mitochondria do not show any significant changes, while the

ER is frequently transformed. Agranular membranes which are often abundant and in close contact with glycogen in precancerous storage cells are usually transformed into ergastoplasm by the addition of ribosomes. Similar phenomena have been previously described for the recovery of hepatocytes after starvation, toxic liver injury and hepatectomy (FAWCETT, 1955; ROUILLER, 1957; JEZEQUEL, 1958; THOENES, 1962; BARTOK and VIRAGH, 1965). It is not clear, however, whether or not the changes of the ER during such "regenerative" processes are directly comparable to those observed during carcinogenesis. In the latter case one often finds unusual combinations of the smooth and rough ER, probably intermediary stages of the ER-transformation, which to our knowledge have never been observed under different conditions. One may interpret them formally as more or less dense complexes between multiple "ergastoplasm pockets" and smooth membranes, the latter being associated with glycogen (BANNASCH, 1967 c). Typically, these ergastoplasm pockets enclose glycogen-free islands within zones of accumulated glycogen. The islands contain large amounts of free ribosomes or, sometimes, mitochondria or microbodies. Within small areas of the cytoplasm the formation of such pockets thus leads to a state which is characteristic for the entire cytoplasm of the final hepatoma cell. Here, a lack of glycogen and agranular membranes is always observed together with an unusual abundance of free ribosomes or ergastoplasm. We consider, therefore, the ergastoplasm pockets as an early indication of cellular transformation.

The morphological observations enumerated above probably express a basic change in cellular metabolism. This is supported by the fact that epithelia with enhanced glycogen storage partially or totally lose their capacity to synthesize glycogen in the course of cellular transformation. This process seems to be irreversible, since hepatoma cells are characterized throughout their lifetime by a lack of glycogen, regardless of whether or not the carcinogen still acts upon them. Two observations proved that glycogen depletion is a consequence of the metabolism of hepatoma cells and not caused by a lack of available glucose. First, histochemical methods revealed the presence of precancerous storage cells or liver cells with normal glycogen content in the immediate vicinity of hepatomas. Secondly, biochemical evidence proved in vitro that sections of hepatomas placed in glucose-containing media do not synthesize glycogen, as opposed to normal liver tissue (Novikoff hepatoma: ASHMORE et al., 1958; Morris hepatoma 5123: WEBER et al., 1961).

The other decisive morphological phenomena of cellular transformation, increase in ribosomes and rearrangement of the ER, are probably caused by the same changes in the metabolism of storage cells as is glycogen reduction. Not only do they occur at exactly the same time, they are equally irreversible. Since the transformation of agranular ER into ergastoplasm coincides with glycogen depletion, the direct or indirect functional relationship between the smooth ER and glycogen, discussed in the last chapter, seems to receive further support. In addition, the characteristic changes in shape of the ER during carcinogenesis prove the close relationship between the granular and the agranular ER. Thus, the idea that the smooth ER, synthesized itself by the ergastoplasm (DALLNER et al., 1966), may serve as a reservoir for further ergastoplasm (FAWCETT, 1955; ROUILLER, 1957; JEZEQUEL, 1958; THOENES, 1962; BARTOK and VIRAGH, 1965) is consistent with our findings. This function of the smooth ER, however, is probably prominent only under certain metabolic conditions. During carcinogenesis, the smooth membranes seem to be transformed into ergastoplasm only after the role they play in the storage cell has become superfluous, due to a shift in the metabolism during the metamorphosis of the precancerous cell to the hepatoma cell.

Morphological evidence does not lead us to understand the reason for the metabolic changes and their character. *One is inclined, however, to relate the reduction of the initially abundant glycogen and the concurrent cytoplasmic alterations to the appearance of a WARBURG-type glycolysis* (WARBURG, 1923; 1926; 1966) *during cellular metamorphosis.* Measurement of the metabolic functions during hepatocarcinogenesis could easily be interpreted in these terms. Thus, many authors observed an increase in glycolysis only towards the end of the precancerous phase, that is, at the beginning of tumor growth (ORR and STICKLAND, 1941; DICKENS and WEIL-MALHERBE, 1943; DRUCKREY et al., 1958; FIALA and FIALA, 1959; SYDOW, 1966). It should be mentioned in this context that BURK and coworkers (1965, 1967) recently rejected the statement supported mainly by AISENBERG (1961) and van POTTER (1964) that glycolysis is not a general phenomenon of hepatoma cells or cancer cells. BURK and coworkers studied a variety of transplantable hepatomas and found that all of them showed signs of aerobic glycolysis, its intensity being higher the higher the growth rate.

The reasons for aerobic glycolysis in cancer cells and its significance for carcinogenesis have long been a point of discussion. WARBURG (1966) considers it to be the "prime cause of cancer" ("letzte Ursache des Krebses") and suggests that this special feature of the cancer cell is caused by a primary damage of cellular respiration. The basic difficulties with this hypothesis have already been outlined when the mitochondria, the main site of cellular respiration, were discussed. We saw then that a primary lesion of the mitochondria can probably be ruled out as the reason for carcinogenesis in the liver parenchyma. This is based on evidence presented by DRUCKREY et al., (1958) and HEISE and GÖRLICH (1964) who measured the metabolism of carcinogen-intoxicated liver cells and could not detect any decline in respiration throughout the precancerous phase. Since the typical glycolytic metabolism still developed in both cases, one has to look for different reasons to explain its origin. In view of the present results, an impairment of the carbohydrate metabolism which manifests itself in the form of glycogenosis (glycogen storage disease) must be seriously considered as the reason for the glycolytic metabolism of hepatomas. In this sense, the carcinogen-induced glycogenosis would be answered by an adaptation of cellular enzymes which gradually could redirect the carbohydrate metabolism towards glycolysis and the pentose-phosphate-cycle. This would also be consistent with the fact that the activity of hexokinase, a glucose-phosphorylating enzyme, gradually increases during hepatocarcinogenesis (SHARMA et al., 1965; SYDOW, 1966; SYDOW and SYDOW, 1966). The activity of this enzyme was found to be the limiting factor for the intensity of glycolysis in hepatomas. The activation of glucose-6-phosphate dehydrogenase in most hepatomas (WEBER et al., 1964; MCLEAN and BROWN, 1966) would also be in agreement with the suggested hypothesis. It seems to be indicated, therefore, to re-evaluate the relationship between carcinogen-induced glycogenosis and glycolysis of hepatoma cells. Furthermore it should be investigated whether or not tumors of other parenchymal organs are also preceded by a focal glycogen storage disease, since there are some indications (BANNASCH and SCHACHT, 1968) that this might be a more general phenomenon.

Summary

The present investigations were intended to elucidate the cytogenesis of cancer. Nitrosomorpholine-intoxicated rat liver was used as a model. In comparative studies of several experimental series the development of hepatomas was followed under the light and electron microscope. The main concern was directed at changes in the cytoplasmic fine structure. The most important results are summarized in the following paragraphs (see also text-figure 1, page 64):

1. The action of the carcinogen leads to the formation of two basically different cytotoxic patterns. Depending on their location within the liver lobule we call them "acinuscentral" and "acinusperipheral".

2. The central pattern is characterized by a very pronounced depletion or even total loss of the cytoplasmic glycogen depots, accompanied by a disorganization of the ergastoplasm ("chromatolysis"). Changes in the mitochondria of such epithelia are not uniform. In some cells they retain their normal aspect for many weeks after the beginning of the carcinogenic treatment, in others they exhibit marked structural alterations, such as loss of cristae, separation of the inner compartment by transverse partitions, widening of the interleaflet space of the envelope, loss of intramitochondrial granulae and, in some cases, prolongation and parallel stacking of the cristae. In addition, one can often observe a pronounced enlargement of the chondriom.

3. The loss of the glycogen and ergastoplasm disorganization (acinuscentral cytotoxic pattern) are correlated with the concentration of the carcinogen (single dose). They are usually reversible after removal of the carcinogenic stimulus. They must, therefore, be attributed to a "concentration action" of the carcinogen, in the sense of DRUCKREY and KÜPFMÜLLER. As such they indicate a nonspecific cytotoxic effect. In consistency, one cannot correlate such epithelia with the formation of tumors, if one gradually follows the cytogenesis and histogenesis of hepatomas. Most of the cells of the acinuscentral toxic pattern undergo coagulation necrosis, thereby leading to liver fibrosis and cirrhosis. We consider the acinuscentral pattern, therefore, as a regressive or cirrhogenic parenchymal lesion which does not play any significant role in the formation of hepatomas.

4. The acinusperipheral cytotoxic pattern, on the other hand, indicates a precancerous reaction of the cell. It is mainly characterized by enhanced glycogen storage. The higher glycogen content leads to an enlargement of the cell. Other components of the cytoplasm, and this holds especially true for the basophilic bodies, are pushed towards the periphery of the cell or the vicinity of the cell nucleus. The amount of basophilic bodies (ergastoplasm) per unit volume of the cytoplasm is markedly reduced. The fine structure of the ergastoplasm, however, remains almost invariably unchanged. Thus, one can only observe a "dislocation" and relative reduction of the ergastoplasm in the precancerous storage cells, while a disorganization of the rough ER is typical of the acinuscentral pattern. In some of the storage cells one also observes a hypertrophy of the smooth ER which is usually accompanied by a reduction of the accumulated glycogen. The smooth membrane complexes appear in various formes: first, as the typical network of tubular elements, secondly, as a "coiled network" and finally, as a wavy or concentrically stacked "complex of lamellar cisternae". The smooth membranes are usually closely associated with glycogen particles. The mitochondria of storage cells mostly retain their normal structure, in

rare cases they undergo the same changes as in the epithelia of the acinuscentral pattern. Frequently, glycogen-containing cytolysosomes are present.

5. As opposed to the changes observed in cells of the acinuscentral pattern, glycogen storage and the cytoplasmic alterations associated with it proceed in the course of chronic intoxication. Their intensity and the size of the affected areas parallel the total uptake of the carcinogen. Multiple parenchymal foci with enhanced glycogen storage are thus formed at the end of the precancerous phase. Such foci persist for weeks and months, even after discontinuation of the carcinogen treatment. The carcinogen thus induces a focal glycogenosis (glycogen storage disease), probably as a consequence of irreversible enzyme deficiencies (glucose-6-phophatase? phophorylase?), caused by the toxic substance.

6. Hepatomas always originate from parenchymal foci with enhanced glycogen storage. The decisive step towards the hepatoma manifests itself morphologically by a reduction of the accumulated glycogen and a concomitant increase in the number of ribosomes ("chromatogenesis"), sometimes accompanied by a transitory fatty infiltration of the cytoplasm. In the course of these events, the smooth ER which is often abundant in the storage cells is usually transformed into ergastoplasm. In many storage cells one temporarily finds unusual combinations of smooth and rough ER towards the end of the precancerous phase. They may be formally interpreted as a more or less dense complex of "ergastoplasm pockets" and smooth membranes, the latter being closely associated with glycogen. The endpoint of the transformation of a storage cell into a hepatoma cell is characterized by a definite lack of fat, glycogen and smooth ER together with an unusually high content of ribosomes or ergastoplasm.

7. The transformation of a storage cell into a hepatoma cell is irreversible. Further investigations are necessary to reveal whether or not there exists any relationship between the cellular transformation and the beginning of a WARBURG-type glycolysis. The latter might be due to an adaptation of the carcinogen-intoxicated liver cell to the glycogenosis (glycogen storage disease) which is detectable during the precancerous phase. The investigation of this hypothesis seems to be justified by the present study, especially since it has proved once more that the two frequently discussed theories of carcinogenesis, relating it to a primary lesion of the mitochondria or the ergastoplasm, are still not sufficiently documented.

References

AISENBERG, A. C.: The glycolysis and respiration of tumors. New York-London: Academic Press 1961.

ALBERT, Z., and M. ORLOWSKI: Some peculiar biological and biochemical properties of a mouse hepatoma induced by chrysoidin. I. Biological and biochemical characteristics of the hepatoma. J. nat. Cancer Inst. 25, 443—447 (1960).

ALLARD, C., G. de LAMIRANDE, and A. CANTERO: Enzymes and cytological studies in rat hepatoma transplants, primary liver tumors and in the liver following azo dye feeding or partial hepatectomy. Cancer Res. 17, 862—879 (1957).

ALTMANN, H.-W.: Die Pathologie der Leber. Naturforschung u. Medizin in Deutschland 1939—1946. 72, Spezielle Pathologie, Teil I, 83—141. Weinheim/Bergstraße: Verlag Chemie GmbH 1948.

— Morphologische Pathologie des Cytoplasmas. Die Pathobiosen. In: Handb. d. allg. Path. (Hrsg. BÜCHNER, F., E. LETTERER u. F. ROULET) Bd. II/1, S. 420—612. Berlin-Göttingen-Heidelberg: Springer 1955.

ALTMANN, H. W., u. U. OSTERLAND: Über cytoplasmatische Wirbelbildungen in den Leberzellen der Ratte bei chronischer Thioacetamidvergiftung. Beitr. path. Anat. 124, 1—18 (1960).

ANDERVONT, H. B., and TH. B. DUNN: Transplantation of spontaneous and induced hepatomas in inbred mice. J. nat. Cancer Inst. 13, 455—504 (1952).

— and TH. B. DUNN: Transplantation of hepatomas in mice. J. nat. Cancer Inst. 15, 1513—1526 (1955).

ANTHONY, D. D., F. SCHAFFNER, and F. HUTTERER: Morphological and cytochemical correlation of rat liver cell organelles in ethionine intoxication and modifications. Fed. Proc. 20, 288 (1961).

ARHELGER, R. B., J. S. BROOM, and R. K. BOLER: Ultrastructural hepatic alterations following tannic acid administration to rabbits. Amer. J. Path. 46, 409—434 (1965).

ASHMORE, J., G. WEBER, and B. R. LANDAU: Isotope studies on the pathways of glucose-6-phosphate metabolism in Novikoff hepatoma. Cancer Res. 18, 974—979 (1958).

— — Biological chemistry of glucose-6-phosphatase. In: Vitamins and hormones 17, 91—132 (1959).

ASHWORTH, C. T., E. SANDERS, and N. ARNOLD: Hepatic lipids. Fine structural changes in liver cells after high-fat, high-cholesterol and choline deficient diets in rats. Arch. Path. 72, 625—647 (1961).

—, D. J. WERNER, M. D. GLASS, and N. J. ARNOLD: Spectrum of fine structural changes in hepatocellular injury due to thioacetamide. Amer. J. Path. 47, 917—952 (1965).

BALOGH, K., and S. J. ROTH: Histochemical and electron microscopic studies of eosinophilic granular cells (Oncocytes) in tumors of the parotid gland. Lab. Invest. 14, 310—320 (1965).

BANNASCH, P.: Glykogen und endoplasmatisches Retikulum der Leberzelle während der Carcinogenese. Ber. phys. med. Ges. Würzburg 73, 83—86 (1967 a).

— Nitrosamin-induzierte Glykogenose und Geschwulstbildung in der Rattenleber. Verh. dtsch. Ges. Path. 51, 343—349 (1967 b).

— Ungewöhnliche Formationen des endoplasmatischen Retikulums bei experimenteller Glykogenose der Leber. 13. Tgg. dtsch. Ges. Elektronenmikr. Marburg (1967 c). Mikroskopie (im Druck).

–, u. H. A. MÜLLER: Lichtmikroskopische Untersuchungen über die Wirkung von N-Nitrosomorpholin auf die Leber von Ratte und Maus. Arzneimittel-Forsch. 14, 805—814 (1964).

—, u. U. SCHACHT: Nitrosamin-induzierte tubuläre Glykogenspeicherung und Geschwulstbildung in der Rattenniere. Virch. Arch. Abt. B Zellpath. 1, 95—97 (1968).

BARKA, T., P. J. SCHEUER, F. SCHAFFNER, and H. POPPER: Structural changes of liver cells in copper intoxication. Arch. Path. 78, 331—349 (1966).

BARTOK, J., u. ST. VIRAGH: Zur Entwicklung und Differenzierung des endoplasmatischen Retikulums in den Epithelzellen der regenerierenden Leber. Z. Zellforsch. 68, 741—754 (1965).

BASSI, M.: Electron microscopy of rat liver after carbon tetrachloride poisoning. Exp. Cell Res. 20, 313—323 (1960).

BAUDHUIN, P., H. G. HERS, and H. LOEB: An electron microscopic and biochemical study of type II glycogenosis. Lab. Invest. 13, 1139—1152 (1964).

BENEDETTI, E. L., and P. EMMELOT: Changes in the fine structure of rat liver cells brought about by dimethylnitrosamine. Proc. Europ. Reg. Conf. on Electron. Microsc. Delft, 1960, Delft 1961, Vol II, 875—878.

— — Effect of dimethylnitrosamine on the endoplasmic reticulum of rat liver cells. Lab. Invest. 15, 126—209 (1966).

BERNHARD, W., F. HAGUENAU, A. GAUTIER et CH. OBERLING: La structure submicroscopique des éléments basophiles cytoplasmique dans le foie, le pancréas et les glandes salivaires. Z. Zellforsch. 37, 282—300 (1952).

—, and C. ROUILLER: Close topographical relationship between mitochondria and ergastoplasma of liver cells in a definite phase of cellular activity. J. biophys. biochem. Cytol. 2 (suppl.) 73—78 (1956).

BIBERFELD, P., J. L. E. ERICSSON, P. PERLMANN, and M. RAFFTELL: Ultrastructural features of in vitro propagated rat liver cells. Z. Zellforsch. 71, 153—168 (1966).

BÖRNIG, H., A. HORN u. V. MÜCKE: Das Verhalten der Glucose-6-Phosphatase in der normalen und geschädigten Leber. Acta biol. med. germ. 9, 623—631 (1962).

BROCK, N., H. DRUCKREY u. H. HAMPERL: Die Erzeugung von Leberkrebs durch den Farbstoff 4-Dimethylaminoazobenzol. Z. Krebsforsch. 50, 431—455 (1940).

BRODEHL, J.: Thioacetamid in der experimentellen Leberforschung. Klin. Wschr. 39, 956—962 (1963).

BROUWERS, J. A. J., and P. EMMELOT: Microsomal N-demethylation and the effect of the hepatic carcinogen dimethylnitrosamine on amino-acid incorporation into the proteins of rat livers and hepatomas. Exp. Cell Res. 19, 467—474 (1960).

BRUNI, C.: Hyaline degeneration of rat liver cells studied with electron microscope. Lab. Invest. 9, 209—215 (1960).

— Observations on the fine structure of slow and fastgrowing rat hepatomas. Fifth Internat. Congr. f. Electr. Microsc. (ed. S. S. BREESE) Vol 2, PP-13 (1962).

—, and K. R. PORTER: The fine structure of the parenchymal cell of the normal rat liver. Amer. J. Path. 46, 691—755 (1965).

BÜCHNER, F.: Die experimentelle Kanzerisierung der Parenchymzelle in der Synopsis klassischer und moderner morphologischer Methoden. Verh. dtsch. Ges. Path. 45, 37—59 (1961).

BURK, D., and M. WOODS: Newer aspects of glucose fermentation in cancer growth and control. Arch. Geschwulstforsch. 28, 305—319 (1967).

—, M. WOODS, and J. HUNTER: On the cancer metabolism of minimal deviation hepatomas. Proc. Amer. Ass. Cancer Res. 6, 9 (1965).

CARRUTHERS, J. S., and J. W. STEINER: Experimental extrahepatic biliary obstruction. Fine structural changes of liver cell mitochondria. Gastroenterology 42, 419—430 (1962).

CHANG, J. P., J. D. SPAIN, and A. C. GRIFFIN: Histochemical manifestation of early changes in rat liver during carcinogenesis induced by 3-methyl-4-dimethylaminoazobenzene. Cancer Res. 18, 670—675 (1958).

CHIESARA, E., F. CLEMENTI, F. CONTI, and J. MELDOLESI: The induction of drug-metabolizing enzymes in rat liver during growth and regeneration. A biochemical and ultrastructural study. Lab. Invest. 16, 254—267 (1967).

CLEMENTI, P.: Variazioni biochimiche e ultrastrutturali epatiche durante carenza proteica. Atti Accad. med. lombarda 15, 405—421 (1960).

COIMBRA, A., and S. P. LEBLOND: Sites of glycogen synthesis in rat liver cells as shown by electron microscope radioautography after administration of glucose-H³. J. Cell Biol. 30, 151—175 (1966).

CONFER, D. B., and R. J. STENGER: Nodules in the livers of C3H mice after long-term carbon tetrachloride administration: a light and electron microscopic study. Cancer Res. 26, 834—843 (1966).

COSSEL, L.: Die menschliche Leber im Elektronenmikroskop. Jena: VEB Gustav Fischer 1964.

CÔTÉ, J., W. OEHLERT u. F. BÜCHNER: Autoradiographische Untersuchungen zur DNS-Synthese während der experimentellen Kanzerisierung der Leberparenchymzelle der Ratte durch Diäthylnitrosamin. Beitr. path. Anat. 127, 450—473 (1962).

DALLNER, G., PH. SIEKEVITZ, and G. E. PALADE: Biogenesis of endoplasmic reticulum membranes. I. Structural and chemical differentiation in developing rat hepatocyte. J. Cell. Biol. 30, 73—96 (1966 a).

— — — Biogenesis of endoplasmic reticulum membranes. J. Cell. Biol. 30, 97—117 (1966 b).

DALTON, A. J.: An electron microscopical study of a series of chemically induced hepatomas. In: Cellular controle mechanisms and cancer (ed. P. EMMELOT and O. MÜHLBOCK) p. 211—225. Amsterdam-London-New York: Elsevier Publishing Company 1964.

DAOUST, R.: Cellular populations and nucleic acid metabolism in rat liver parenchyma during azo dye carcinogenesis. Canad. Cancer Conf. 5, 225—239 (1963).

—, and F. MOLNAR: Cellular populations and mitotic activity in rat liver parenchyma during azo dye carcinogenesis. Cancer Res. 24, 1898—1909 (1964).

DAVID, H : Submikroskopische Ortho- und Pathomorphologie der Leber. Berlin: Akademie-Verlag 1964.

—, u. L.-H. KETTLER: Degeneration von Lebermitochondrien nach Ammoniumintoxikation. Z. Zellforsch. 53, 857—866 (1961).

DeDuve: The lysosome concept. In: Lysosomes (ed. A. V. S. de Reuck and M. P. Cameron) p. 1—31. London: J. & A. Churchill Ltd. 1963.

—, and P. Baudhuin: Peroxisomes (microbodies and related particles). Physiol. Rev. 46, 323—357 (1966).

Dickens, F., and H. Weil-Malherbe: The metabolism of normal and tumor tissue. XX. A comparison of the metabolism of tumors of liver and skin with that of the tissue of origin. Cancer Res. 3, 73—87 (1943).

Djaczenko, W., and Z. Albert: Ultrastructure of chrysoidin—induced hepatoma. Acta med. pol. 3, 307—321 (1962).

Driessens, J., A. DuPont et A. Demaille: L'hépatome expérimental azoïque du rat examiné au microscope électronique. C. R. Soc. Biol. 153, 788—790 (1959).

Drochmans, D.: Mise en évidence du glycogène dans les cellules hépatiques par microscopie électronique. J. biophys. biochem. Cytol. 8, 553—558 (1960).

— Morphologie du glycogène. J. Ultrastruct. Res. 6, 141—163 (1962).

Druckrey, H., F. Bresciani u. H. Schneider: Atmung und Glykolyse der Rattenleber während der Behandlung mit 4-Dimethylaminoazobenzol. Z. Naturforsch. 13 b, 514 (1958).

—, P. Dannenberg, W. Dischler u. D. Steinhoff: Reinzucht von 10 Rattenstämmen (BD-Stämmen) und Analyse des genetischen Pigmentierungssystems. Arzneimittelforsch. 12, 911—919 (1962).

—, u. K. Küpfmüller: Dosis und Wirkung. Aulendorf (Württbg): Ed. Cantor 1949.

—, R. Preussmann, S. Ivancovic u. D. Schmähl: Organotrope carcinogene Wirkungen bei 65 verschiedenen N-Nitrosoverbindungen an BD-Ratten. Z. Krebsforsch. 69, 103—201 (1967).

— —, D. Schmähl u. M. Müller: Chemische Konstitution und carcinogene Wirkung bei Nitrosaminen. Naturwissenschaften 48, 134—145 (1961).

—, D. Schmähl, W. Dischler u. A. Schildbach: Quantitative Analyse der experimentellen Krebserzeugung. Naturwissenschaften 49, 217—228 (1962).

Edwards, J. E., and J. White: Pathologic changes with special reference to pigmentation and classification of hepatic tumors in rats fed p-dimethylaminoazobenzene (butter yellow). J. nat. Cancer Inst. 2, 157—183 (1941/42).

Emmelot, P., and E. L. Benedetti: Changes in the fine structure of rat liver cells brought about by dimethylnitrosamine. J. biophys. biochem. Cytol. 7, 393—395 (1960).

—, I. J. Mizrahi, R. Naccaroto, and E. L. Benedetti: Changes in function and structure of the endoplasmic reticulum of rat liver cells after administration of cysteine. J. biophys. biochem. Cytol. 12, 177—180 (1962).

Epstein, Sh., N. Ito, L. Merkow, and E. Farber: Cellular analysis of liver carcinogenesis: the induction of large hyperplastic nodules in the liver with 2-fluorenylacetamide or ethionine and some aspects of their morphology and glycogen metabolism. Cancer Res. 27, 1702—1711 (1967).

Ericsson, J. L. E., and St. Orrenius: Evolution of phenobarbital induced alterations in the endoplasmic reticulum of hepatic parenchymal cells. J. Ultrastruct. Res. 14, 418—419 (1966).

— — and I. Holm: Alterations in canine liver cells induced by protein deficiency. Ultrastructural and biochemical observations. Exp. mol. Path. 5, 329—349 (1966).

—, and B. F. Trump: Electron microscopic studies of the epithelium of the proximal tubule of the rat kidney. I. The intracellular localization of acid phosphatase. Lab. Invest. 13, 1427—1456 (1964).

Essner, E.: Endoplasmic reticulum and the origin of microbodies in the fetal mouse liver. Lab. Invest. 17, 71—87 (1967).

—, and A. B. Novikoff: Cytological studies on two functional hepatomas. Interrelations of endoplasmic reticulum, Golgi apparatus and lysosomes. J. Cell. Biol. 15, 289—312 (1962).

Farber, E.: Similarities in the sequence of early histological changes induced in the liver of the rat by ethionine, 2-acetylamino-fluorene and 3-methyl-4-dimethylaminoazobenzene. Cancer Res. 16, 142—148 (1956).

— Ethionine Carcinogenesis. Advanc. Cancer Res. 7, 383—474 (1963).

Fasske, E., u. H. Themann: Die elektronenmikroskopische Struktur menschlicher Carcinome. Beitr. path. Anat. 122, 313—344 (1960).

Fawcett, D. W.: Observations on the cytology and electron microscopy of hepatic cells. J. nat. Cancer Inst. 15, 1475—1503 (1955).

FAWCETT, D. W.: Structural and functional variations in the membranes of the cytoplasm. In: Intracellular Membraneous Structure (ed. S. SENO, and E. V. COWDRY) p. 15—40. Okayama: Japan Society of Cell Biology 1965.

—, and J. W. WILSON: A note on the occurence of virus-like particles in the spontaneous hepatomas of C3H mice. J. nat. Cancer Inst. 15, 1505—1511 (1955).

FIALA, S., and A. E. FIALA: On the correlation between metabolic and structural changes during carcinogenesis in rat liver. Brit. J. Cancer 13, 136—151 (1959).

FIRMINGER, H. J.: Histopathology of carcinogenesis and tumors of the liver in rats. J. nat. Cancer Inst. 15, 1417—1442 (1954).

FOUTS, J. R.: Interaction of drugs and hepatic microsomes. Fed. Proc. 21, 1107—1111 (1962).

FURUTA, Y.: Electron microscopic studies on the fine structure of the liver. II. Hepatoma induced by dimethylaminoazobenzene in rats. Jap. Arch. int. Med. 3, 820—830 (1956).

GAUTIER, A.: Technique de coloration de tissus dans des polyester. Experientia (Basel) 16, 124 (1960).

GHADIALLY, F. N., and E. W. PARRY: Ultrastructure of a human hepatocellular carcinoma and surrounding non-neoplastic liver. Int. J. Cancer 19, 1989—2004 (1966).

GIRBARDT, M.: Lebendnachweis von Einzelelementen des endoplasmatischen Retikulum. J. Cell Biol. 27, 343—440 (1965).

GÖSSNER, W., u. H. FRIEDRICH-FREKSA: Histochemische Untersuchungen über die Glucose-6-Phosphatase in der Rattenleber während der Cancerisierung durch Nitrosamine. Z. Naturforsch. 19 b, 862—864 (1964).

GORANSON, E. S., J. McBRIDE, and G. WEBER: Phosphorylase activity in rat hepatoma and mouse mammary carcinoma transplants. Cancer Res. 14, 227—231 (1954).

GRAFFI, A., u. H. BIELKA: Probleme der experimentellen Krebsforschung. Leipzig: Akademische Verlagsgesellschaft, Geest und Portig K. G. 1959.

—, u. W. HEBEKERL: Über chemische Frühveränderungen der Rattenleber nach Verfütterung cancerogener Azofarbstoffe. Arch. Geschwulstforsch. 5, 1—24 (1953).

GREENAWALT, J. W., G. V. FOSTER, and A. L. LEHNINGER: The observation of unusual membranous structures associated with liver mitochondria in thyrotoxic rats. In: Electron microscopy. Proc. Fifth Intern. Congr. Electron Microscop. Philadelphia, II, p. 00-5, New York-London: Academic Press 1962.

GRISHAM, J. W.: Early changes in the fine structure of the hepatic cell in ethionine-fed rats. Fed. Proc. 19, 186 (1960).

— Electron microscope study of carcinoma of the liver in ethionine-fed rats. Gastroenterology 38, 792—793 (1960).

GRUNDMANN, E.: Die Zytogenese des Krebses. Dtsch. med. Wschr. 86, 1077—1084 (1961).

— Die Krebsentwicklung als intrazelluläres Problem dargestellt am Diäthylnitrosaminkrebs der Rattenleber. Mitteilungsdienst GBK 2, 589—633 (1962).

— Über intrazelluläre Vorgänge während der Carcinogenese. Münch. med. Wschr. 34, 1662—1667 (1966).

—, u. H. SIEBURG: Die Histogenese und Cytogenese des Lebercarcinoms der Ratte durch Diäthylnitrosamin im lichtmikroskopischen Bild. Beitr. path. Anat. 126, 57—90 (1962).

GUPTA, D. N.: Acute changes in the liver after administration of thioacetamide. J. Path. Bact. 72, 183—192 (1956).

GUSTAFSSON, R. G., and B. A. AFZELIUS: Comparative effects on rat liver cells after dimethylnitrosamine, 2-fluorenamine or prednisolone treatment studied by electron microscopy. J. nat. Cancer Inst. 30, 1045—1075 (1963).

HADJIOLOV, A. A., and K. I. DANCHEVA: Phosphorylase activity of primary rat liver cancer. Nature 181, 547—548 (1958).

HAGUENAU, F.: The ergastoplasm. Its history, ultrastructure and biochemistry. Int. Rev. Cytol. 7, 425—438 (1958).

HARTMANN, H. A.: Rat liver after N-hydroxy-2-acetylaminofluorene. Ultrastructure. Arch. Path. 79, 126—134 (1965).

HARTROFT, W. ST.: Some electron microscopic features of the liver in experimental choline deficiency. In: Aktuelle Probleme der Hepatologie (ed. G. A. MARTINI) p. 53—57. Stuttgart: Georg Thieme 1962.

Hauss, W. H., H. J. Albrecht, U. Gerlach, G. Junge-Hülsing, J. Ch. Kingreen, S. Ritter, H. Themann u. G. Wüst: Untersuchungen während der Kanzerisierungsphase durch Diäthylnitrosamin. Mitteilungsdienst GBK 3, 323—340 (1964).

Heine, U., A. Graffi, H. J. Helmcke u. A. Randt: Elektronenmikroskopische Untersuchungen an normalen Zellen und Tumorzellen der Rattenleber. Z. ärztl. Fortbild. 51, 648—652 (1957).

Heinlein, H., K. Hübner, K. J. Lennartz u. G. Rudolph: Neuere Erkenntnisse zur Geschwulstbildung in der Leber. Klin. Wschr. 40, 121—125 (1962).

Heise, E., u. M. Görlich: Veränderungen der Stoffwechselgrößen und der Aktivitäten glykolytischer Fermente während der Kanzerisierung von Rattenlebern durch Diäthylnitrosamin. Exp. Cell Res. 33, 289—300 (1964).

Herdson, P. B., P. J. Garvin, and R. B. Jennings: Reversible biological and fine structural changes produced in rat liver by a thiohydantoin compound. Lab. Invest. 13, 1014—1031 (1964 a).

— — — Fine structural changes in rat liver induced by phenobarbital. Lab. Invest. 13, 1032—1037 (1964 b).

— — — Fine structural changes produced in rat liver by partial starvation. Amer. J. Path. 45, 157—182 (1964 c).

Herman, L., L. Eber, and P. J. Fitzgerald: Liver cell degeneration with ethionine administration. In: Electron microscopy. Fifth Intern. Congr. Electr. Microscop. Philadelphia (ed. S. S. Breese) Vol. 2, p. VV—6 (1962).

Hers, H. G.: α-glucosidase deficiency in generalyzed glycogen storage disease (Pompe's disease). Biochem. J. 86, 11—21 (1963).

— Glycogen storage disease. Advanc. metabol. disorders 1, 2—44 (1964).

Hobik, H. P., u. E. Grundmann: Quantitative Veränderungen der DNS und RNS in der Rattenleberzelle während der Carcinogenese durch Diäthylnitrosamin. Beitr. path. Anat. 127, 25—48 (1962).

Holzmann, K., u. R. Lange: Zur Zytologie der Glandula parathyreoidea des Menschen. Weitere Untersuchungen an Epithelkörperchenadenomen. Z. Zellforsch. 58, 759—789 (1963).

Howatson, A. F., and A. W. Ham: Electron microscope study of sections of two rat liver tumors. Cancer Res. 15, 62—69 (1955).

Hruban, Z., W. H. Kirsten, and A. Slesers: Fine structure of spontaneous hepatic tumors of C3H/fGs mice. Lab. Invest. 15, 576—588 (1966).

—, B. Spargo, H. Swift, R. W. Wissler, and R. G. Kleinfeld: Focal cytoplasmic degredation. Amer. J. Path. 42, 657—683 (1963).

—, H. Swift, F. W. Dunn, and D. E. Lewis: Effect of β-3-furylalanine on the ultrastructure of the hepatocytes and pancreatic acinar cells. Lab. Invest. 14, 70—80 (1965).

— — and M. Rechcigl: Fine structure of transplantable hepatomas of the rat. J. nat. Cancer Inst. 35, 459—473 (1965).

— — and R. W. Wissler: Alterations in the fine structure of hepatocytes produced by β-3-thienylalanine. J. Ultrastruct. Res. 8, 236—250 (1963).

Hübner, G.: Die Kanzerogenese in der Rattenleber nach Verfütterung von Buttergelb. Fifth Intern. Congr. Electr. Microscop. Philadelphia (ed. S. S. Breese). Vol. 2, p. PP-6 (1962).

—, u. W. Bernhard: Das submikroskopische Bild der Leberzelle nach temporärer Durchblutungssperre. Beitr. path. Anat. 125, 1—30 (1961).

—, F. Paulussen u. O. Kleinsasser: Menschliche Epithelzellen mit spontaner Hyperplasie abartiger Mitochondrien. Die Onkozyten. 13. Tgg. dtsch. Ges. Elektronenmikr., Marburg (1967). Mikroskopie (im Druck).

Hultin, T., E. Arrhenius, H. Löw, and P. N. Magee: Toxic liver injury. Inhibition by dimethylnitrosamine of incorporation of labeled amino acids into proteins of rat liver preparations in vitro. Biochem. J. 76, 109 (1960).

Hunter, N. W.: Experimental liver poisoning of the frog (rana pipiens) with carbon tetrachloride. I. Liver-glycogen storage, blood glucose level and cytochrome-oxidase activity. Exp. mol. Path. 4, 449—455 (1965).

Ito, N., and E. Farber: Effects of trypan blue on hepatocarcinogenesis in rats given ethionine or N-2-fluorenylacetamid. J. nat. Cancer Inst. 37, 775—783 (1966).

JÉZÉQUEL, A.-M.: Les effects de l'intoxication aigue au phosphore sur le foie de rat. Etude au microscope électronique. Ann. anat. path. 3, 512—537 (1958).
— Dégenérescence myélinique des mitochondries de foie humain dans épithélioma du cholédoque et un ictère viral. J. Ultrastruct. Res. 3, 210—215 (1959).
—, K. ARAKAWA, and W. STEINER: The fine structure of the normal, neonatal mouse liver. Lab. Invest. 14, 1894—1930 (1965).
JONES, A. L., D. T. ARMSTRONG, and D. W. FAWCETT: Increased cholesterol biosynthesis following phenobarbital induced hypertrophy of agranular endoplasmic reticulum in liver. Proc. Soc. exp. Biol. Med. 119, 1136—1139 (1965).
—, and D. W. FAWCETT: Hypertrophy of the agranular endoplasmic reticulum in hamster liver induced by phenobarbital (with a review of the functions of this organelle in liver). J. Histochem. Cytochem. 14, 215—232 (1966).
JORDAN, S. W.: Electron microscopy of hepatic regeneration. Exp. mol. Path. 3, 183—200 (1964).
KARNOVSKY, M. J.: Simple methods for "staining with lead" at high pH in electron microscopy. J. biophys. biochem. Cytol. 11, 729—732 (1961).
— The fine structure of mitochondria in the frog nephron correlated with cytochrome oxidase activity. Exp. mol. Path. 2, 347—366 (1963).
KELLY, M. G., R. W. O'GARA, H. ADAMSON, K. GADEKAR, D. C. BOTKIN, W. H. REESE, and W. T. KERBER: Induction of hepatic cell carcinomas in monkeys with N-nitrosodiethylamine. J. nat. Cancer Inst. 36, 323—351 (1966).
KENDREY, G., J. JUHASZ u. S. BELA: Mit Isonikotinsäurehydrazid hervorgerufene Leberveränderungen und Lebertumoren. Elektronenmikroskopische Untersuchungen an weißen Ratten. Morph. és Jg. Orv. Szemle 7, 176—186 (1967).
KIERNAN, F.: The anatomy and physiology of the liver. Phil. Trans. Roy. Soc. London 123, 711—770 (1833).
KINOSITA, R.: Studies on the cancerogenic chemical substances. Trans. Soc. Path. Jap. 27, 665—725 (1937).
KRAMSCH, D., V. BECK u. W. OEHLERT: Einfluß der Äthioninvergiftung und Nahrungsentzuges auf die DNS-Neubildung in den Wechselgeweben und parenchymatösen Organen der Ratte. Beitr. path. Anat. 128, 416—444 (1963).
KRÖGER, H., u. B. GREUER: Einfluß des carcinogenen N-nitrosomorpholins auf die Induzierbarkeit von Tryptophan-Oxygenase und Tyrosin-2-Oxoglutarat-Transaminase in der Rattenleber. Hoppe-Seyler's Z. physiol. Chem. 342, 148—155 (1965).
KURTZ, S. M.: A new method for embedding tissues in Vestopal W. J. Ultrastruct. Res. 5, 468 (1963).
LAFONTAINE, J. G., and C. ALLARD: A light and electron microscope study of the morphological changes induced in rat liver cells by the azo dye 2-ME-DAB. J. Cell Biol. 22, 143—172 (1964).
LANE, B. P., and CH. S. LIEBER: Effects of butylated hydrotoluene on the ultrastructure of rat hepatocytes. Lab. Invest. 16, 342—348 (1967).
LANGE, R.: Zur Histologie und Cytologie der Glandula parathyreoidea des Menschen. Z. Zellforsch. 53, 765—828 (1961).
LANGER, K. H.: Vergleichende histologische und funktionelle Untersuchungen an Niere, Leber und Herzmuskel nach experimentellen Blutverlusten und orthostatischen Kollapsen am Kaninchen. Inaug. Diss. Würzburg 1965.
LEAK, L. V., J. B. CAULFIELD, J. F. BURKE, and C. F. McKHANN: Electron microscopic studies on a human fibromyxosarcoma. Cancer Res. 27, 261—285 (1967).
—, and V. J. ROSEN: Early ultrastructural alterations in proximal tubular cells after unilateral nephrectomy and X-irradiation. J. Ultrastruct. Res. 15, 326—348 (1966).
LE BRETON, E., and Y. MOULÉ: Biochemistry and physiology of the cancer cell. In: The cell (ed. J. BRACHET, and A. E. MIRSKY). Bd. V, 497—544 (1961).
LEJEUNE, N., D. THINES-SEMPOUX, and G. HERS: Tissue fractionation studies. 16. intracellular distribution and properties of α-glycosidase in rat liver. Biochem. J. 86, 16 (1963).
LUBARSCH, O.: Über die Bedeutung der pathologischen Glykogenablagerungen. Virch. Arch. 183, 188—228 (1906).

Luck, D. J. L.: Glycogen synthesis from uridine diphosphate glucose. The distribution of the enzyme in liver cell fractions. J. biophys. biochem. Cytol. 10, 195—209 (1961).

Luft, R., K. Ikkos, G. Palmieri, L. Ernster, and B. Afzelius: A case of severe hypermetabolism of non-thyroid origin with a defect in the maintenance of mitochondrial respiratory control. A correlated clinical biochemical and morphological study. J. clin. Invest. 41, 1776—1804 (1962).

Luse, S. A., and A. Mikata: An electron microscopic study of hepatic changes produced by N-2-fluorenyldiacetamide. Fed. Proc. 22, 483 (1963).

Ma, M. H., and A. J. Webber: Fine structure of liver tumors induced in the rat by 3'-methyl-4-dimethylaminoazobenzene. Cancer Res. 26, 935—946 (1966).

Magee, P. N.: Toxic liver injury. Inhibition of uptake of labeled aminoacids into liver protein in vivo by dimethylnitrosamine. Biochem. J. 65, 31 P (1957).

— Toxic liver injury. Inhibition of protein synthesis in rat liver by dimethylnitrosamine. Biochem. J. 70, 606—611 (1958).

— Toxic liver necrosis. Lab. Invest. 15, 111—130 (1966).

—, and J. M. Barnes: The production of malignant primary hepatic tumors in the rat by feeding dimethylnitrosamine. Brit. J. Cancer 10, 114—122 (1956).

—, and E. Farber: Toxic liver injury and carcinogenesis. Methylation of rat liver nucleic acids by dimethylnitrosamine in vivo. Biochem. J. 83, 114—124 (1962).

—, and T. Hultin: Toxic liver injury and carcinogenesis. Methylation of proteins of rat liver slices by dimethylnitrosamine in vitro. Biochem. J. 83, 106—113 (1962).

Mao, P., and J. J. Molnar: The fine structure and histochemistry of lead-induced renal tumors in rats. Amer. J. Path. 50, 571—603 (1967).

Man, J. C. H. de: Observations, with the aid of the electron microscope, on the mitochondrial structure of experimental liver tumors in the rat. J. nat. Cancer Inst. 24, 795—819 (1960).

— Effect of cortisone on the fine structure, glycogen content, and glucose-6-phosphatase activity of hepatic cells in fasted and dimethylnitrosamine-treated rats. Cancer Res. 24, 1347—1361 (1964).

—, and A. P. R. Block: Relationship between glycogen and agranular endoplasmic reticulum in hepatic cells. J. Histochem. Cytochem. 14, 135—146 (1966).

McGavran, M. H.: The ultrastructure of papillary cystadenoma lymphomatosum of the parotid gland. Virch. Arch. path. Anat. 338, 195—202 (1965).

McLean, P., and J. Brown: Activities of some enzymes concerned with citrate and glucose metabolism in transplanted rat hepatomas. Biochem. J. 98, 874—882 (1966).

Merkow, L. P., Sh. M. Epstein, B. J. Caito, and B. Bartus: The cellular analysis of liver carcinogenesis: ultrastructural alterations within hyperplastic liver nodules induced by 2-fluorenylacetamide. Cancer Res. 27, 1712—1721 (1967).

— —, R. C. Gauer, and E. Farber: Ultrastructural alterations within hyperplastic liver nodules induced by 2-acetylaminofluorene. J. Cell Biol. 31, 76 A (1967).

Meyer-Bertenrath, J.: Alterationen der Ribosomenstruktur während der durch Nitrosomorpholin induzierten Carcinogenese. Hoppe-Seyler's Z. physiol. Chem. 348, 645—650 (1967 a).

— Störungen der Ribosomenfunktion während einer exogen induzierten Carcinogenese. Z. Naturforsch. 22 b, 660—662 (1967 b).

—, u. U. Dege: Zur Wirkung des Nitrosomorpholins auf den RNS-Stoffwechsel der Rattenleber. Z. Naturforsch. 22 b, 169—172 (1967).

Mikata, A., and S. A. Luse: Ultrastructural changes in the rat liver produced by N-2 fluorenyl-diacetamide. Amer. J. Path. 44, 455—480 (1964).

Miller, F.: Orthologie und Pathologie der Zelle im elektronenmikroskopischen Bild. Verh. dtsch. Ges. Path. 42, 261—332 (1958).

Miller, J. A., and E. C. Miller: The carcinogenic aminoazo dyes. Advanc. Cancer Res. 1, 340—396 (1953).

Millonig, G., and K. R. Porter: Structural elements of rat liver cells involved in glycogen metabolism. Proc. Europ. Region Congr. Electr. Microsc. Delft, 655—659 (1960).

Miyaji, H.: Zit. nach H. P. Morris. Advanc. Cancer Res. 9, 227 (1965).

MÖLBERT, E.: Das elektronenmikroskopische Bild der Leberparenchymzelle nach histotoxischer Hypoxidose. Beitr. path. Anat. **118**, 203—227 (1957).
—, K. HILL u. F. BÜCHNER: Die Kanzerisierung der Leberparenchymzelle durch Diäthylnitrosamin im elektronenmikroskopischen Bild. Beitr. path. Anat. **126**, 218—242 (1962).
MORRIS, H. P.: Hepatocarcinogenesis by 2-acetylaminofluorene and related compounds including comments on dietry and other influences. J. nat. Cancer Inst. **15**, 1535—1545 (1954).
— Studies on the development, biochemistry and biology of experimental hepatomas. Advanc. Cancer Res. **9**, 228—296 (1965).
MÜLLER, M.: Autoradiographische, fluoreszenzimmunologische und morphologische Untersuchungen zur Verteilung und Wirkung von o-Amidoazotoluol (OAAT) in der Mäuseleber. Arch. Geschwulstforsch. **30**, 97—108 (1967).
MUKHERJEE, T., R. G. GUSTAFSSON, B. A. AFZELIUS and E. ARRHENIUS: Effects of carcinogenic amines on amino acid incorporation by liver slices. Cancer Res. **23**, 944—953 (1963).
MULAY, A. S., and H. J. FIRMINGER: Liver tumors induced in rats by p-dimethylaminoazobenzene-1-Azo-1-Naphthalene compared with tumors induced by p-dimethylaminoazobenzene. J. nat. Cancer Inst. **13**, 35—55 (1952).
—, and R. W. O'GARA: Hepatic lesions in rats fed azo dye in laboratory chow. Arch. Path. **81**, 162—165 (1966).
NIGAM, V. N.: Glycogen metabolism in liver during DAB carcinogenesis. Brit. J. Cancer **19**, 912—919 (1965).
NOVIKOFF, A. B.: A transplantable rat liver tumor induced by 4-dimethylaminoazobenzene. Cancer Res. **17**, 1010—1027 (1957).
— Mitochondria (Chondriosomes). In: The cell (ed. J. BRACHET and A. E. MIRSKY), Bd. II, 299—421. New York-London: Academic Press 1961.
— Lysosomes and related particles. In: The cell (ed. J. BRACHET and A. E. MIRSKY), Bd. II, 423—488. New York-London: Academic Press 1961.
OBERLING, CH., and W. BERNHARD: The morphology of the cancer cell. In: The cell (ed. J. BRACHET, and A. E. MIRSKY), Bd. V, 405—496. New York-London: Academic Press 1961.
—, et CH. ROUILLER: Les effets de l'intoxication aigue au tétrachlorure de carbone sur le foie du rat. Etude au microscope électronique. Ann. Anat. Path. **1**, 401—427 (1956).
OEHLERT, W., u. J. HARTJE: Die Veränderungen des Eiweiß- und Ribonucleinsäurestoffwechsels während der experimentellen Kanzerisierung durch Diäthylnitrosamin. Beitr. path. Anat. **128**, 376—415 (1963).
OPIE, E. L.: The influence of diet on the production of tumors of the liver by butter yellow. J. exp. Med. **80**, 219—230 (1944).
— The pathogenesis of tumors of the liver produced by butter yellow. J. exp. Med. **80**, 231—246 (1944).
— Mobilisation of basophile substance (ribonucleic acid) in the cytoplasm of liver cells with the production of tumors by butter yellow. J. exp. Med. **84**, 91—106 (1946).
ORR, J. W., and D. E. PRICE: Observations on the hepatotoxic action of the carcinogen p-dimethylaminoazobenzene. Path. Bact. **60**, 461—469 (1948).
— —, and L. H. STICKLAND: The glycogen content of rats' liver after poisoning with large doses of p-dimethylaminoazobenzene. J. Path. Bact. **60**, 573—581 (1948).
—, and L. H. STICKLAND: The metabolism of rat liver during carcinogenesis. Biochem. J. **35**, 479—487 (1941).
ORRENIUS, S., and L. E. ERICSSON: Enzyme-membrane relationship in phenobarbital induction of synthesis of drug-metabolizing enzyme system and proliferation of endoplasmic membranes. J. Cell. Biol. **29**, 181—198 (1966).
ORTEGA, P.: Light and electron microscopy of rat liver after feeding with DDT. Fed. Proc. **21**, 306 (1962).
— Light and electron microscopy of dichlorodiphenyl-trichloroethane (DDT) poisoning in the rat liver. Lab. Invest. **15**, 657—679 (1966).
PALADE, G. E.: Studies on the endoplasmic reticulum. II. Simple disposition in cells in situ. J. Biophys. biochem. Cytol. **1**, 567—582 (1955).
—, and PH. SIEKEVITZ: Liver microsomes: an integrated morphological and biochemical study. J. biophys. biochem. Cytol. **2**, 171—200 (1956).

PARKS, H. F.: Unusual formations of ergastoplasm in parotid acinous cells of mice. J. Cell Biol. 14, 221—234 (1962).

PETERS, V. B., H. M. DEMBRITZER, C. W. KELLY, and E. BARUCH: Ergastoplasmic changes associated with glycogenolysis. Fifth Intern. Congr. Electr. Microsc. (ed. S. S. BREESE), vol 2 TT-7. New York. Academic Press 1962.

PHILLIPS, M. J., N. J. UNAKAR, G. DOORNEWAARD, and J. W. STEINER: Glycogen depletion in the newborn rat liver. An electron microscopic and electronhistochemic study. J. Ultrastruct. Res. 18, 142—165 (1967).

PITOT, H. C., and Y. S. CHO: Control mechanisms in the normal and neoplastic cell. Prog. exp. Tumor. Res. 7, 158—223 New York-Basel: Karger 1965.

—, and C. PERAINO: Endoplasmic reticulum and hepatocarcinogenesis. Biochemical clinics. Vol 3: The liver, p. 139. New York: The Reuben H. Donnelley Corporation 1964.

PORTER, K. R.: The ground substance; observations from electron microscopy. In: The cell (ed. J. BRACHET and A. E. MIRSKY), pp. 621—675. New York-London: Academic Press 1961.

—, and C. BRUNI: An electron microscope study of the early effects of 3'-Methyl-DAB on rat liver cells. Cancer Res. 19, 997—1009 (1959).

— — Fine structural changes in rat liver cells associated with glycogenesis and glycogenolysis. Anat. Rec. 136, 260—261 (1960).

— — Comparative fine structure of slow and fast growing hepatomas. Acta Un. int. Cancr. 20, 1271—1272 (1964).

POTTER, V. R.: Biochemical studies on minimal deviation hepatomas. In: Cellular control mechanism and cancer (ed. P. EMMELOT and O. MÜHLBOCK), p. 190—210. Amsterdam-London-New York: Elsevier Publishing Company 1964.

PRICE, J. M., E. C. MILLER, J. A. MILLER, and G. M. WEBER: Studies on the intracellular composition of livers from rats fed various aminoazo dyes. I. 4-aminoazobenzene, 4-dimethylaminoazobenzene, 4'-methyl- and 3'-methyl-4-dimethylaminoazobenzene. Cancer Res. 9, 398—402 (1949).

— — — — Studies on the intracellular composition of livers from rats fed various aminoazo dyes. II. 3'methyl-, 2'methyl, and 2-methyl-4-dimethylaminoazobenzene, and 4'-fluoro-4-dimethylaminoazobenzene. Cancer Res. 10, 18—27 (1950).

QUINN, P. S., and J. HIGGINSON: Reversible and irreversible changes in experimental cirrhosis. Amer. J. Path. 47, 353—370 (1965).

RAJEWSKY, M. F., W. DAUBER, and H. FRANKENBERG: Liver carcinogenesis by diethylnitrosamin in the rat. Science 152, 83—85 (1966).

RAPPAPORT, A. M.: Acinar units and the pathophysiology of the liver. In: The liver (ed. CH. ROUILLER), 1, p. 265—328. New York-London: Academic Press 1963.

REMMER, H.: Detoxification of drugs in the liver. In: Progress in liver diseases (ed. H. POPPER and F. SCHAFFNER), 2, 116—133. New York and London: Grune & Stratton 1965.

—, u. H.-J. MERKER: Enzyminduktion und Vermehrung von endoplasmatischem Retikulum in der Leberzelle während der Behandlung mit Phenobarbital (Luminal). Klin. Wschr. 41, 276—283 (1963).

— — Effect of drugs on the formation of smooth endoplasmic reticulum and drug-metabolizing enzymes. Ann. N. Y. Acad. Sci. 123, 79—97 (1965).

REUBER, M. D.: Development of preneoplastic and neoplastic lesions of the liver in male rats given 0,025 percent n-2-fluorenyldiacetamide. J. nat. Cancer Inst. 34, 697—724 (1965).

— Histopathology of transplantable hepatic carcinomas induced by chemical carcinogens in rat. Gann, Monograph, 1, 43—55 (1966).

REVEL, J. P.: Electron microscopy of glycogen. J. Histochem. Cytochem. 12, 104—114 (1964).

—, D. W. FAWCETT, and C. W. PHILPOTT: Observations on mitochondrial structures. Angular configurations of the cristae. J. Cell Biol. 16, 187—195 (1963).

REYNOLDS, E. S.: Liver parenchymal cell injury. I. Initial alterations of the cell following poisoning with carbon tetrachloride. J. Cell Biol. 19, 139—157 (1963).

ROODYN, D. B.: The mitochondrion. In: Enzyme cytology (ed. D. B. ROODYN), p. 103—100. New York-London: Academic Press 1967.

ROUILLER, CH.: Contribution de la microscopie électronique à l'étude du foie normal et pathologique. Ann. Anat. Path. 2, 548—562 (1957).

ROUILLER, C. H.: Physiological and pathological changes in mitochondria morphology. Int. Rev. Cytol. **9**, 227—292 (1960).

—, and A. M. JÉZÉQUEL: Electron microscopy of the liver. In: The liver (ed. CH. ROUILLER), **1**, 195—252. New York-London: Academic Press 1963.

—, et G. SIMON: Contribution de la microscopie électronique au progrès de nos connaissance en cytologie et en histologie hépatique. Rev. intern. hépatol. **12**, 167—206 (1962).

RYTER, A., et E. KELLENBERGER: L'inclusion au polyester par l'ultramicrotomie. J. Ultrastruct. Res. **2**, 200—214 (1958).

SALOMON, J. C.: Modifications ultrastructurale des hepatocytes au cours de l'intoxication chronique par la thioacetamide. In: Electron microscopy. Fifth Intern. Congr. Electr. Microsc. Philadelphia (ed. S. S. BREESE), Vol. 2, p. VV 7, (1962 a).

— Modifications des cellules du parenchyme hépatique du rat sous l'effet de la thioacétamide. J. Ultrastruct. Res. **7**, 293—307 (1962 b).

—, et A. M. JÉZÉQUEL: Ultrastructure des hépatomes experimenteaux. Les tumeurs malignes du foie. Paris: Masson et Cie 1963.

—, M. SALOMON et W. BERNHARD: Modification des cellules du parenchyme hépatique du rat sous l'effet de la thioacetamide. Bull. Cancer **49**, 139—158 (1962).

SASAKI, T., u. T. YOSHIDA: Experimentelle Erzeugung des Leberkarzinoms durch Fütterung mit o-Amidoazotoluol. Virch. Arch. **95**, 175—200 (1935).

SCARPELLI, D. G., M. H. GREIDER, and W. J. FRAJOLA: Observations on hepatic cell hyperplasia, adenoma and hepatoma of rainbow trout (Salmo gairnerii). Cancer Res. **23**, 848—857 (1963).

SCHAUER, A.: Histochemische und biochemische Untersuchungen bei der Kanzerisierung mit Diäthylnitrosamin an der Rattenleber. Verh. dtsch. Ges. Path. **51**, 344—347 (1966).

SCHMÄHL, D.: Entstehung, Wachstum und Chemotherapie maligner Tumoren. Aulendorf (Württbg.): Ed. Cantor 1963.

—, u. R. PREUSSMANN: Kanzerogene Wirkung von Nitrosodimethylamin bei Ratten. Naturwissenschaften **46**, 175 (1960).

— — u. H. HAMPERL: Leberkrebserzeugende Wirkung von Diäthylnitrosamin nach oraler Gabe bei Ratten. Naturwissenschaften **47**, 89 (1960).

SCHNEIDER, W. C., G. H. HOGEBOOM, E. SHELTON, and H. J. STRIEBICH: Enzymatic and chemical studies on the livers and liver mitochondria of rats fed 2-methyl- or 3'-methyl-4-dimethylaminoazobenzene. Cancer Res. **13**, 285—288 (1953).

SCHRAMM, T., H. BIELKA u. A. GRAFFI: Geschwulsterzeugung durch chemische Substanzen. In: Handb. d. exp. Pharmakologie (ed. O. EICHLER, H. HERKEN und A. D. WELCH), Bd. XVI/12, p. 1—241, Berlin, Heidelberg, New York: Springer 1966.

SELJELID, R., and L. E. ERICSSON: An electron microscopic study of mitochondria in renal clear cell carcinoma. J. Microscopie **4**, 759—770 (1965).

SHARMA, R. M., C. SHARMA, A. J. DONELLY, H. P. MORRIS, and S. WEINHOUSE: Glucose-ATP phosphotransferases during hepatocarcinogenesis. Cancer. Res. **25**, 193—199 (1965).

SHIPKEY, F. H., PH. H. LIEBERMANN, F. W. FOOTE and F. W. STEWART: Ultrastructure of alveolar soft part sarcoma. Cancer **17**, 821—830 (1964).

SICKINGER, K., R. KATTERMANN u. H. HANNEMANN: Zunahme von Lebergewicht und Leberglykogen unter Infusion von Cholinorotat und Adenosin. Acta hepato-splenol. (Stuttg.) **14**, 88—99 (1967).

SIE, H.-G., and A. HABLANIAN: Depletion of glycogen synthetase and increase of glucose-6-phosphat dehydrogenase in livers of ethionine-treated mice. Biochem. J. **97**, 32—36 (1965).

SIMARD, A., and R. DAOUST: DNA synthesis and neoplastic transformation in rat liver parenchyma. Cancer Res. **26**, 1665—1672 (1966).

SMUCKLER, E. A., O. A. ISERI, and E. P. BENDITT: Studies on carbon tetrachloride intoxication. I. The effect of carbon tetrachloride on incorporation of labelled amino acids into plasma proteins. Biochem. Biophys. Res. Commun. **5**, 270 (1961).

— — — An intracellular defect by carbon tetrachloride. J. exp. Med. **116**, 55—72 (1962).

SOOSTMEYER, TH.: Glykogengehalt und Zellstrukturen der Leber während des anaphylaktischen Schocks. Virch. Arch. **306**, 554—569 (1940).

SPAIN, J.: Precancerous metabolic alterations in the process of azo dye carcinogenesis. Texas Rep. Biol. Med. **14**, 528—537 (1956).

Spain, J., and A. C. Griffin: A histochemical study of glycogen alterations in the livers of rats following azo dye administration. Cancer Res. 17, 200—204 (1957).

Stanton, M. F.: Diethylnitrosamine-induced hepatic degeneration and neoplasia in the aquarium fish brachydanio rerio. J. nat. Cancer Inst. 34, 117—130 (1965).

Steiner, J. W., and C. M. Baglio: Electron microscopy of the cytoplasm of parenchymal liver cells in α-naphthylisothiocyanate-induced cirrhosis. Lab. Invest. 12, 765—790 (1963).

—, K. Miyai, and M. J. Phillips: Electron microscopy of membrane-particle arrays in liver cells of ethionine-intoxicated rats. Amer. J. Path. 44, 169—213 (1964).

—, M. J. Phillips, and K. Miyai: Ultrastructural and subcellular pathology of the liver. Int. Rev. Cytol. 3, 65—167 (1964).

Stenger, R. J.: Hepatic parenchymal cell alterations after long term carbon tetrachloride administration. A light and electron microscopic study. Amer. J. Path. 43, 867—895 (1963).

— Concentric lamellar formations in hepatic parenchymal cells of carbon tetrachloride-treated rats. J. Ultrastruct. Res. 14, 240—253 (1966).

Stewart, H. L., and K. C. Snell: The histopathology of experimental tumors of the liver of the rat. In: The physiopathology of cancer (ed. F. Homburger and W. H. Fischmann), IInd edit. New York: Paul B. Hoeber 1959.

Stöcker, E.: Der Proliferationsmodus in Niere und Leber. Verh. dtsch. Ges. Path. 50, 53—74 (1966).

Striebich, M. J., E. Shelton, and W. C. Schneider: Quantitative morphological studies on the livers and liver homogenates of rats fed 2-methyl- or 3'-methyl-4-dimethylaminoazobenzene. Cancer Res. 13, 279—284 (1953).

Strong, L. C., et G. M. Smith: Preuve histologique de la présence de glycogène dans un hepatoma malin transplantable. Bull. Ass. franç. Cancer 26, 694—698 (1937).

Svoboda, D. J.: Fine structure of hepatomas induced in rats with p-dimethylaminoazobenzol. J. nat. Cancer Inst. 33, 315—339 (1964).

—, and D. L. Azarnoff: Response of hepatic microbodies to a hypolipidemic agent, ethyl chlorophenoxyisobutyrate (CPIB). J. Cell Biol. 30, 442—450 (1966).

—, G. Grady, and J. Higginson: The effect of chronic protein deficiency in rats. II. Biochemical and ultrastructural changes. Lab. Invest. 15, 731—749 (1966).

—, and J. Higginson: Ultrastructural changes produced by protein and related deficiencies in the rat liver. Amer. J. Path. 45, 353—380 (1964).

Sydow, G.: Beziehungen zwischen Hexokinaseaktivität und Glykolyse in Rattenleber und Hepatomen. Z. Naturforsch. 21 b, 232—237 (1966).

—, und H. Sydow: Die Wirkung kurzdauernder Verabreichung von kanzerogenen Stoffen auf den Glykogengehalt sowie auf die Hexokinase- und Glukokinaseaktivität der Rattenleber. Acta biol. med. germ. 14, 468—475 (1965 a).

— — Die Wirkung von Diäthylnitrosamin bei unterschiedlicher Dosierung auf die Hexokinase- und Glukokinaseaktivität sowie den Glykogengehalt der Rattenleber. Acta biol. med. germ. 15, 20—29 (1965 b).

— — Die Wirkung einmaliger bzw. langfristiger Applikation von 3'-Methyl-4-dimethylaminoazobenzol auf einige Stoffwechselfunktionen der Rattenleber. Arch. Geschwulstforsch. 28, 341—346 (1966).

Tandler, B., and F. H. Shipkey: Ultrastructure of Warthin's Tumor. J. Ultrastruct. Res. 11, 292—305 (1964).

Themann, H.: Elektronenoptische Untersuchungen über das Glykogen im Zellstoffwechsel. Veröffentl. morphol. Path. 66, Stuttgart: Gustav Fischer 1963.

Theron, J. J.: Acute Liver injury in ducklings as a result of aflatoxin poisoning. Lab. Invest. 14, 1586—1603 (1965).

—, and R. C. P. M. Mekel: Electron microscopical studies of human malignant hepatoma cells. T. Gastro-ent. 7 b, 152—164 (1964).

Thoenes, W.: Giemsa-Färbung an Geweben nach Einbettung in Polyester („Vestopal") und Methacrylat. Z. wiss. Mikr. 64, 406 (1960).

— Zur Kenntnis des glatten Endoplasmatischen Retikulums der Leberzelle. Verh. dtsch. Ges. Path. 46, 202—206 (1962).

THOENES, W.: Mikromorphologie des Nephron nach temporärer Ischämie. Zwanglose Abhand-lungen auf dem Gebiet der normalen u. path. Anatomie (Hrsg. W. BARGMANN u. W. DOERR), H. 15. Stuttgart: Georg Thieme 1964.

—, u. P. BANNASCH: Elektronen- und lichtmikroskopische Untersuchungen am Cytoplasma der Leberzellen nach akuter und chronischer Thioacetamid-Vergiftung. Virch. Arch. 335, 556—583 (1962).

THOMAS, C.: Zur Morphologie der durch Diäthylnitrosamin erzeugten Leberveränderungen und Tumoren bei der Ratte. Z. Krebsforsch. 64, 224—233 (1961).

TIMME, A. H., and L. G. FOWLE: Effects of p-dimethylaminoazobenzene on the fine structure of rat liver cells. Nature 200, 694—695 (1963).

TOKER, C., and N. TREVINO: Ultrastructure of human primary hepatic carcinoma. Cancer 19, 1594—1606 (1966).

TROTTER, N. L.: Electron microscopic observations on cytoplasmic components of trans-plantable hepatomas in mice. J. nat. Cancer Inst. 30, 113—133 (1963).

TUJIMURA, H.: Electron microscopic studies in the DAB (p-dimethylaminoazobenzene) -in-duced hepatoma. Med. J. Osaka Univ. 9, 135—145 (1958).

— Electron microscopic studies on the DAB (p-dimethylaminoazobenzene)-induced hepa-toma cells. Med. J. Osaka Univ. 9, 147—161 (1958).

WARBURG, O.: Über den Stoffwechsel der Tumoren. Berlin: Julius Springer 1926.

— Weiterentwicklung der Zellphysiologischen Methoden. Stuttgart: Georg Thieme 1962.

— Über die Ursache des Krebses. In: Molekulare Biologie des malignen Wachstums (Hrsg. H. HOLZER u. A. W. HOLLDORF), p. 1—16. Berlin-Heidelberg-New York: Springer 1966.

WATSON, M. L.: Staining of tissue sections for electron microscopy with heavy metals. J. bio-phys. biochem. Cytol. 8, 727—729 (1958).

WEBER, G.: Behavior of liver enzymes in hepatocarcinogenesis. Advanc. Cancer Res. 6, 403—494 (1961).

—, and A. CANTERO: Glucose-6-phosphatase activity in normal, precancerous and neoplastic tissues. Cancer Res. 15, 105—108 (1955).

—, M. C. HENREY, S. R. WAGLE, and D. S. WAGLE: Correlations of enzyme activities and metabolic pathways with growth rate of hepatomas. Advanc. Enzyme Regulation 2, 335—346 (1964).

—, H. P. MORRIS, W. C. LOVE, and J. ASHMORE: Comparative biochemistry of hepatomas. II. Isotope studies of carbohydrate metabolism in Morris hepatoma 5123. Cancer Res. 21, 1406—1411 (1961).

WILSON, J. W., and E. H. LEDUC: Mitochondrial changes in the liver of essential fatty acid-deficient mice. J. Cell Biol. 16, 281—296 (1963).

WOLFARTH-BOTTERMANN, K. E.: Morphologische Aspekte der Mitochondrienvermehrung. In: Funktionelle und morphologische Organisation der Zelle. III. Probleme der biologischen Reduplikation (ed. H. SITTE), p. 189—313. Berlin- Heidelberg-New York: Springer 1966.

WOOD, R. L.: The fine structure of hepatic cells in chronic ethionine poisoning and during recovery. Amer. J. Path. 46, 307—330 (1965).

YOSHIDA, T.: Zit. nach SASAKI u. YOSHIDA. Virch. Arch. path. Anat. 295, 175 (1935).

Subject Index

Herstellung: Konrad Triltsch, Graphischer Betrieb, Würzburg

Monographs already Published

BOIRON, M., Paris: The Viruses of the Leukemia-sarcoma Complex

CAVALIERE, R., A. ROSSI-FANELLI, B. MONDOVI, and G. MORICCA, Roma: Selective Heat Sensitivity of Cancer Cells

CHIAPPA, S., Milano: Endolymphatic Radiotherapy in Malignant Lymphomas

DENOIX, P., Villejuif: Le traitement des cancers du sein

FISHER, E. R., Pittsburgh: Ultrastructure of Human Normal and Neoplastic Prostate

FUCHS, W. A., Bern: Lymphography in Cancer

GRUNDMANN, E., Wuppertal-Elberfeld: Morphologie und Cytochemie der Carcinogenese

IRLIN, I. S., Moskva: Mechanisms of Viral Carcinogenesis

LANGLEY, F. A., and A. C. CROMPTON, Manchester: Epithelial Abnormalities of the Cervix Uteri

MATHÉ, G., Villejuif: L'Immunotherapie des Cancers

MEEK, E. S., Bristol: Antiviral and Antitumour Agents of Biological Origin

MULLER, J. H., Zürich: Therapeutic Incorporations of Radiopharmaceuticals in Cancer and Allied Diseases

NEWMAN, M. K., Detroit: Neuropathies and Myopathies Associated with Occult Malignancies

OGAWA, K., Osaka: Ultrastructural Enzyme Cytochemistry of Azo-dye Carcinogenesis

PACK, G. T., New York: Clinical Aspects of Cancer Immunity and Cancer Susceptibility

PACK, G. T., and A. H. ISLAMI, New York: Tumors of the Liver

PARKER, J. W., Los Angeles: Lymphocyte Transformation

RITZMAN, S. E., and W. C. LEVIN, Galveston: The Syndrome of Macroglobulinemia

ROY-BURMAN, P., Los Angeles: Biochemical Mechanisms Involved in the Inhibition of Metabolic Processes by Purine, Pyrimidine, and Nucleoside Analogs

SUGIMURA, T., Tokyo, H. ENDO, Fukuoka, and T. ONO, Tokyo: Chemistry and Biological Action of 4-Nitroquinoline 1-oxide, a Carcinogen

SZYMENDERA, J., Warsaw: The Metabolism of Bone Mineral in Malignancy

WEIL, R., Lausanne: Biological and Structural Properties of Polyoma Virus and its DNA

WILLIAMS, D. C., Caterham, Surrey: The Basis for Therapy of Hormon Sensitive Tumours

WILLIAMS, D. C., Caterham, Surrey: The Biochemistry of Metastasis